M000228753

High-Throughput Satellites

For a listing of recent titles in the
Artech House Space Technology and Applications Series,
turn to the back of this book.

High-Throughput Satellites

Hector Fenech

ARTECH
HOUSE

BOSTON | LONDON
artechhouse.com

Library of Congress Cataloging-in-Publication Data
A catalog record for this book is available from the U.S. Library of Congress

British Library Cataloguing in Publication Data
A catalog record for this book is available from the British Library.

ISBN-13: 978-1-63081-825-8

Cover design by Kalyani Kastor

© **2021 Artech House**
685 Canton St.
Norwood, MA

10 9 8 7 6 5 4 3 2 1

To my Father who passed on his passion for communications.

To Patricia for her support while writing this book.

To all engineers pushing the envelope of HTS.

Contents

Preface

Satellite communications date back to the 1950s when the first artificial satellites were launched. The first of these, Sputnik 1, launched in 1957, was more of an atmospheric research satellite. In 1958, with limited success, NASA launched the Pioneer 1, its first satellite intended to orbit the moon to relay infrared images. The first satellite to merit designation as a communication satellite is perhaps Score, also launched in 1958. Primitive by today's standards, Score employed the store-and-forward technique using a tape recorder.

Sir Arthur C. Clarke had postulated the concept of geostationary satellites back in 1945 in *Wireless World*, a magazine that I used to read when I was young—although not in 1945! Syncom 2 was the first geosynchronous communication satellite, although it was not quite geostationary as it had an inclination of about 33°. Syncom 2 was followed a year later by Syncom 3, which was the first geostationary communication satellite. It was used to broadcast the 1964 Olympic Games from Tokyo, Japan.

Commercial geostationary satellite communications started in 1965 with the launch of Intelsat's satellite, Intelsat 1, also known as Early Bird. Eutelsat started operations with the cooperation of the European Space Agency in 1982 with its Orbital Test Satellite (OTS). Since then, various satellite applications have been developed, but the one that seems to be gaining the most attention among satellite operators and the space industry is the high-throughput satellite (HTS).

The demand for connectivity—whether at home, in the office, or on the move on land, at sea, or in the sky—is constantly on the rise. HTS is a solution, and the HTS solution is in constant evolution.

This book looks at the services that exploit HTS systems, their respective operational and commercial contexts, and the challenges that come with these specific services. In fact, one of the challenges of defining an HTS system is the evolving environment: operational, technical, and commercial. Accordingly, the book outlines the underlying theory that governs these systems for both the forward and return links and that determines their performance. It surveys the current operational HTS systems, highlighting their respective salient features and some of their tendencies. HTS systems started with geostationary satellites, but today low-Earth orbit (LEO) systems are being planned and deployed, and 5G is round the corner. With these alternative solutions come threats and opportunities, offering more options to customers, who are becoming increasing technology-agnostic. All this pushes satellite operators to rethink their business models and the satellite industry to redesign its products.

It is highly likely that in the future, there will not be a single solution for connectivity and that instead a system of systems may evolve. In this scenario, each component system capitalizes on its performance for the suited application. It is also very probable that the envelope of applicability for systems will be dependent on technology and, therefore, by no means stagnant in time. In such a diverse environment, innovation is the key to survival.

With any satellite, geostationary equatorial orbit (GEO) or LEO, the basic resources remain power and mass. Translating these resources into a sustainable business will lead to success and maximize the market share of satellites to provide better broadband communications to their users. Short-lifetime satellites like those used in LEO, with their associated reliability requirements, have an advantage that may allow them to employ enabling technologies faster.

With GEO, on the other hand, the dream of a digital payload makes the satellites less bespoke and therefore more standardized. It should also make GEO HTSs less expensive and faster to produce. With examples appearing in industry, it seems that these objectives are being realized. There are several techniques under study, and enabling technologies could certainly allow for new payload architectures. Payload architectures will continue to evolve.

My interest in HTSs started with KA-SAT, which was launched in 2010—although work on the project had started some 10 years earlier. It was the first of the Ka-band generation of HTSs. As the system architect, I had to learn fast. It was an exciting and challenging period. While I was not alone, my group within Eutelsat was small, so the work was intensive but rich. There were numerous exchanges with industry players that were considering the construction of the satellite. These discussions were enlightening, and the need to consider a commercial schedule often brought my feet back to the ground.

Other HTSs followed: Eutelsat 36C/AMU 1, Eutelsat 65WA, and Konnect. As the business case, the commercial and operational requirements, the technologies, and the business as a whole evolved, the optimization criteria for the system followed suit—and the learning experience continued.

I am passionate about the subject, because it is an application of a technology to connect people. Until the early 2000s, being on a commercial aircraft meant that one would be disconnected. Similarly, on a ship, connectivity ended once land was out of sight. That is no longer the case. Now we can choose to be connected or not, and there seems to be an insatiable—and ubiquitous—demand for connectivity. Consider, for example, soldiers and humanitarian workers sent to remote areas where a connectivity infrastructure may not exist; satellite systems are essential to provide connectivity to such places. Accordingly, engineers will continue to develop solutions.

There are many excellent books on satellite communications but not many that focus on HTSs. This book aims to fill that void, presenting a consolidated approach to the subject and tackling the issue from a system point of view. There is probably material for several books going into each of the covered aspects (and others not addressed in this book). My aspiration is to share my passion with system engineers in the field, engineers thinking of moving into this area, university students with an interest in this discipline, and perhaps even consultants who would like to get a deeper insight into HTSs.

I tried to make sure that, in addition to providing technical information, the book touches on the commercial and business environments that lead to technical decisions. I believe that if the definition of the system is more holistic, the solution will make more business sense and consequently will be more sustainable with the opportunity to evolve, grow, and better serve its users.

Acknowledgments

I would like to thank my ex-colleagues of the Future Satellite System Group who over the years supported me and indirectly and directly contributed to this book. I also would like to show my gratitude to friends working with satellite operators, and in the space industry for their support.

1

Introduction

This book addresses HTSs, the latest evolution in communication satellites aimed at meeting growing communication demands for smaller terminals that offer more capacity at a lower per-unit cost. Satellite communication is sometimes considered to be an expensive last resort. Fortunately, however, HTSs offer capacities that are orders of magnitude higher than those of traditional satellites and that are optimized for broadband connectivity delivering Internet with costs per megabits per second that are comparable to those of terrestrial systems.

The first commercial communication satellites in the 1960s provided telecommunication services (i.e., television, telephone, and telefacsimile). At that time, the role of satellites in television services was mainly to make contents available to nodes for distribution to viewers. Telephony was analog, and telefacsimile was the data communications of the time. Intelsat 1, also known as Early Bird, provided a transatlantic link in what, today, might fall under the label of trunking. Signals from various sources arrived at an Earth station that was linked to another Earth station via satellite, and then the signals were distributed to their final destinations. Big Earth stations were required, because the satellite resources were limited. C-band was the norm then.

As satellite resources improved, the size of the terminals could be reduced and more specialized services provided. Point-to-point with smaller Earth stations became possible. The very small–aperture terminal (VSAT) services evolved in the C-band with 1-m terminals in the early 1980s and then moved to the Ku-band. VSAT networks, which can be star or meshed, were mainly limited professional services (e.g., corporate networks). It is interesting to note that VSATs used the satellites of the time and therefore often did not use a standard terminal but a range of terminals to adapt to the contours of the single coverage. In this process, the terminal size increases toward the satellite edge of coverage to compensate for the degradation in satellite performance.

In the 1980s, the Ku-band became popular. Interestingly, Eutelsat traces its origins to 1977 but started operations in the Ku-band with the OTS 2 launched in 1978. (OTS 1 was lost at launch.) This was in collaboration with the European Space Agency. The C-band was introduced to Eutelsat in 2002 when Eutelsat acquired a satellite from France Telecom and become known as Eutelsat 5WA. The 1980s also saw the evolution of direct-to-home (DTH) services. The use of the Ku-band together with the introduction of the low-noise gallium arsenide (GaAs) high-electron mobility transistor (HEMT) devices reduced the size of the terminal significantly. Prior to this stage, reception of television at home was possible only if you had a big yard, since the dishes were 2m or more. With the advent of DTH, the size dropped to about 60 cm.

The 1990s and 2000s saw these markets mature. In 2005, VSAT services contributed to 21.4% of the revenue for Eutelsat [1]. The Hot Bird position at 13° east, Eutelsat's prime broadcasting position, grew enormously between the late 1990s and the early 2000s, and today it still carries most of the 6,867 channels broadcast by Eutelsat satellites [2]. For the period 2009–2014, Eutelsat recorded a compound annual growth rate (CAGR) of about 5% in video services [3].

In the 2010s, it started becoming clear that the growth rates of the traditional services could not be sustained indefinitely. The term VSAT disappeared from the vocabulary of Eutelsat reports in 2010 [4]. Although VSATs were still physically present, their importance had diminished. This was probably due to the proliferation of the Internet, which has not ceased since its inception in the United States in the mid 1990s. In the European Union, 30% of households had access to the Internet in 2007. In 2016, that figure had increased to 85% [5]. The Internet was perhaps initially seen as a threat to satellites, and in some ways, it was. Many VSAT networks disappeared. Corporate entities do not need to have their own networks if their sites had access to the Internet.

However, there is no access to the Internet on offshore platforms, in the oceans, or in the sky. In 2019, 90% of households in the EU had access to the Internet, and 73% of individuals had access to the Internet via a mobile phone. What is interesting is that the percentage increase of households with access to the Internet per annum is flattening out. The increase between 2018 and 2019 was 1% [5]. Although the deployment of the fiber infrastructure is expensive, since it serves many nodes, ultimately, the cost per node is attractive. This means that the economic attractiveness of fiber is a function of population density and that reaching out to the increasingly remote part of the population becomes increasingly expensive per node and less economically attractive.

This is where HTSs come in. As the saying goes, "If you can't beat them, join them!" HTSs deliver Internet service to the home using a small terminal (about 70 cm) at an acceptable price. Satellite systems excel at providing coverage over an area, densely populated or not, and therefore can, at least, cater to the clients that the terrestrial infrastructure leaves underserved or even unserved.

The first HTS is considered to be Anik F2, which was launched in 2004 to provide services over Canada, a country with large swaths of sparsely populated areas; it operated in the Ka-band. This was followed in 2005 by IPSTAR 1, also known as Thaicom 4, which operated in the Ku-band and covered parts of the Asia-Pacific region. In 2010, Eutelsat launched KA-SAT to provide Internet service to Europe and parts of the Middle East and North Africa.

The traveling-wave tube amplifier (TWTA), a significant resource on a spacecraft, is an important dimensioning parameter. In KA-SAT, each TWTA delivers 10 times more capacity than a TWTA in a traditional satellite [6], and the introduction of this single satellite tripled the capacity of the Eutelsat fleet at the time. KA-SAT was the first of a generation of Ka-band HTS satellites (i.e., ViaSat 1 and Jupiter 1, also known as EchoStar 17 or XVII) that paved the way to providing Internet access to all citizens, destroying the digital divide. Although technics and technologies continue to evolve, the GEO HTS is now an established class of communications.

The story goes on, because GEO HTSs will no longer be the only satellite solution. There are at least three LEO HTS systems on the drawing board. These projects are even more ambitious and are valued at several billons of euros each—one or two orders of magnitude more expensive than GEO HTS. These systems will expand HTSs to a global reach, making them truly ubiquitous.

There is an overlap in the capabilities of LEO and GEO systems, but there is also some differentiation that may allow them to coexist. The crystal ball gets a bit murky here; only time will reveal the future of the two contenders or complementary systems.

References

[1]   Eutelsat Communications, *Management Report of Consolidated Accounts on 30 June 2006*, https://www.eutelsat.com/files/PDF/investors/2005-06/management-report-consolidated-accounts-0506.pdf.

[2]   Eutelsat Communications, *Third Quarter and Nine Month 2019–20 Revenues*, 14 May 2020, https://www.eutelsat.com/files/PDF/investors/2019-20/Eutelsat_Communications _Q3_2019-20_PR.pdf.

[3]   Fenech, H., "Progress and Evolution of Eutelsat through Responsive System Architectures," Chapter 14 in *Recent Successful Satellite Systems: Visions of the Future*, D. K. Sachdev (ed.), American Institute of Aeronautics and Astronautics, 2017.

[4]   Eutelsat Communications, *Consolidated Financial Statements as of 30 June 2010*, https:// www.eutelsat.com/files/PDF/investors/2009-10/ETL-consolidated-financial-statements 300610.pdf.

[5]   Johnson, J., "Internet Usage in Europe—Statistics & Facts," *Statistica*, 10 February 2020, https://www.statista.com/topics/3853/internet-usage-in-europe/#:~:text=85%20 percent%20of%20European%20households,of%20connected%20households%20 in%202007.

[6]   Fenech, H., "KA-SAT and Future HTS Systems," *The Plenary Session of the 14th IEEE International Vacuum Electronics Conference (IVEC)*, May 21–23, 2013, Paris, France.

2

The Communication Mission Requirements

2.1 Overview

The genesis of a communication satellite is in its commercial and operational requirements. However, the technical genesis stems from the communication mission requirements. These requirements are a technical representation of the commercial and operational requirements.

This chapter discusses the primary considerations of the communication mission requirements, namely the service area, the available spectrum for users, the gateways, the capacity, the effective isotropic radiated power and received figure of merit, and the channelization.

These system parameters are interlinked so that a payload with a given service area would yield more capacity if the power were concentrated over a smaller service area. Similarly, more capacity would be available if more spectrum were available.

2.2 The Service Area

The service area definition defines the formal definition of the geographical area where the service will be delivered. It is an expression of the commercial requirements.

Coverage is used to define the satellite performance, since the coverage is given by the antenna performance, which must at least encompass the service area. Thus, a service area can have shape corners, while the coverage would have curvatures, which are a function of the antenna size.

The service area and the coverage are not independent. There is more leeway for a small service area, but the tendency for HTS systems is to have large service. There is a limit of what part of the Earth a satellite can see from a given geostationary orbital location. Theoretically, an orbital location with a longitude that is close to the center of the service ensures the best elevations toward the parts of the service area that are most remote from the satellite. In the real world, however, choices for the orbital location may be limited, and the service area may have to be tweaked accordingly.

At the system level, the service area can be measured in square degrees as measured from the satellite. The simple example of a service area shown in Figure 2.1 would have an area of $\alpha\beta^{\circ 2}$. In an HTS, the coverage for the user link is a regular contiguous lattice of circular beams.

The service area can be defined as one of the following:

- A list of countries;
- A polygon of geographical coordinates.

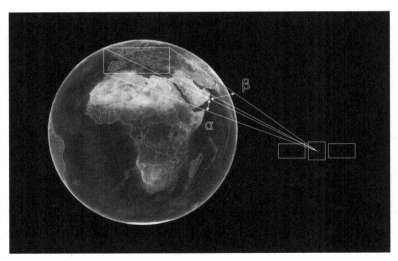

Figure 2.1 An illustration of the definition of a service area in square degrees subtended at the satellite.

An example of the service area definition in terms of countries could include, for example, Albania, Austria, Belgium, Bosnia and Herzegovina, Bulgaria, Croatia, Republic of Cyprus, Czech Republic, Denmark, Estonia, Finland, France, Germany, Greece, Hungary, Iceland, Ireland, Italy, Latvia, Lithuania, Luxembourg, Malta, Netherlands, Norway, Poland, Portugal, Romania, Serbia, Slovakia, Slovenia, Spain, Sweden, Switzerland, and the United Kingdom. Figure 2.2 shows a polygon description of the same service area.

The example in Figure 2.2 demonstrates the possible pitfalls. If we look at the polygon description north of the United Kingdom, the Shetland Islands and the Faroes have been omitted when officially they are part of the United Kingdom and Denmark, respectively. Similar European examples include the Canary Islands, Ceuta, and Melilla for Spain; Lampedusa, Pantelleria, and Linosa for Italy; and the Azores for Portugal. Similar examples of remote islands or areas that are part of a country exist in other parts of the world. The inclusion of such areas would stretch the coverage and possibly require extra hardware; therefore, a trade-off would be required based on the economic and political value.

Figure 2.3 shows a possible implementation for the user service area. Notice in Figure 2.3 that the main coverage is a regular lattice of circular beams. Some leeway is available for isolated areas. Whereas classical satellites tend to have a service area implemented with a single contoured beam, HTSs have the lattice of circular beams a main feature. In broadband access, the bulk

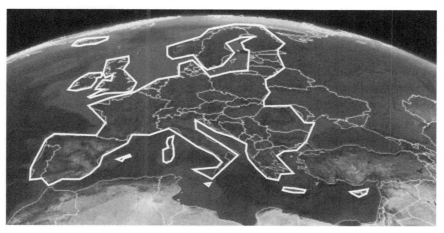

Figure 2.2   An example of the polygon definition of a service area.

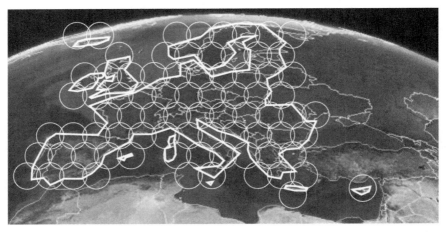

Figure 2.3   A multibeam coverage solution for the polygon service area shown in Figure 2.2.

of the traffic is point-to-point, which means that to make the most optimal use of the space segment resources, the smallest possible beams are required. However, as the beams get smaller the number of the beams required for a given service area increases, and consequently the system complexity increases.

The user links would have an uplink user service area and a downlink user service area for the return link and the forward link, respectively. In general, the geographical definition of both these service areas would be identical. Additionally, there would be an uplink gateway service area and a downlink gateway service area for the forward link and the return link, respectively. Similarly, the geographical definition of both these service areas would be identical and correspond to isolated beams consistent to the gateway locations with the required suppression of mutual interference.

Although the coverage is not officially part of the communication mission requirements, it has a significant bearing on a satellite's design and complexity and on its compliance with requirements. At the first level, it defines the granularity for meeting the service area requirements. It also determines the size of the antennas. Thus, smaller beams would be better for meeting the service area requirements but would necessitate larger antennas. Additionally, for a service area of $\alpha\beta^{\circ 2}$ and a beam area of $k\theta_b^{\circ 2}$, where θ_b° is the beam diameter, the number of beams is given as

$$N = \frac{\alpha\beta}{k\theta_b^2} \qquad (2.1)$$

Thus, decreasing the diameter by a factor will multiply the number of beams by the square of the factor. This has a significant impact on the payload, since it implies a corresponding multiplication of equipment with the associated impact on equipment accommodation, mass, and power.

Observe that most beams are circular (or elliptical), so that $k = \pi/4$. Although this is true, circles do not pack together seamlessly. Accordingly, it is useful to introduce the idea of a cell, which is the part of the beam that is used to cover a portion of the service area. Typically, even though it is covered by a circular beam, the cell is a hexagon that is circumscribed in the circular beam and $k = 3\sqrt{3}/8$, which is less than the previous value of k by 17%.

Along with the coverage comes the associated color scheme(s). A color scheme defines the reutilization of resources so that a given color is associated with a given resource. The most common application of the color scheme is for frequency reuse. In a homogeneous system where all beams have the same capacity and the maximum capacity per beam is sought, a four-color scheme is optimal. Since two orthogonal polarizations are employed, an even number color scheme is implicit. Thus, the ensuing color scheme becomes four colors with each color using half the available bandwidth, with two colors on one polarization and the other two on the orthogonal polarization.

In a nonhomogenous HTS system, where the capacity requirement per beam is not constant, more elaborate schemes are possible, and the requirements for the even number color scheme may disappear. Thus, for example, one polarization may have two colors that may not entail equal bandwidth, and the other polarization may have three colors with unequal bandwidth. The objective here would be to apportion the available spectrum in a mix that corresponds to the capacity requirements of the beams.

Color schemes can also be applied to antennas, so that a color may represent beams sharing the same reflector. This could also be four so that a reflector could be associated to part of the spectrum on a given polarization. However, the association of a reflector to a part of the spectrum and a given polarization is not necessary and three reflectors could be used. The difference in the hardware is obvious but the difference in the performance is that in a three-color scheme, the beams from the same reflector are closer to each other and therefore the intrasystem interference is higher. Four-color antenna schemes are generally preferred for a solely HTS satellite, but in a multimission satellite, the antenna accommodation quickly becomes an issue, and the three-color antenna scheme may become a compelling solution.

The capacity of a HTS system is a function of the geographical distribution of the user terminals. If all the user terminals are concentrated at the peaks of the beams, the system capacity is higher than if the terminals are concentrated at the edges of the beams. Likewise, if there is a higher density of terminals in an area—the western portion of the service area, for example—a situation may arise in which the beams in the higher density zone are saturated, while the beams in rest of the service area are underutilized. The concept of sellable capacity becomes evident, because in order to reap the maximum economic benefit from the system, it is not sufficient to just have the capacity. The capacity must be where it is required. To complicate the issue further, the capacity demand is not static and may evolve over time [1].

It is therefore important to adopt a basis to monitor the system performance. A uniform distribution is a simplest approach, but it is rarely representative of the commercial and operational reality. In the demand map methodology, socioeconomic indicators are used to develop the demand per pixel. In this context, a pixel is the smallest area of the demand map for which the capacity information is available. It determines the resolution of the demand map and, therefore, should be considerably smaller than the area covered by a cell. With a beamwidth of 0.5°, a 20 km-by-20 km pixel size may be adequate. The system can then be optimized using the demand map. Demand maps give a static picture. Although multiple demand maps can be developed to represent the evolution of the capacity demand over time, the accuracy often diminishes with time. This is where flexibility becomes important. Flexibility is apportioning the capacity between the cells so that the highest possible portion of the capacity is sellable over the lifetime of the satellite.

2.3 Spectrum

The available spectrum is the most influential factor on the system performance and economics. Having more spectrum and/or having contiguous spectrum reduces the payload complexity and reduces the per-unit capacity cost, especially for the space segment. It is important to remember that the amount of spectrum available for a given application is determined by the Radio Regulations [2] of the International Telecommunications Union (ITU), which is responsible for establishing the segmentation of the spectrum and the usage of these parts of the spectrum at the international level.

The ITU splits the world in three regions, listed as follows.

- Region 1: Europe (including Iceland), Africa, the Middle East, Armenia, Azerbaijan, the Russian Federation, Georgia, Kazakhstan, Mongolia, Uzbekistan, Kyrgyzstan, Tajikistan, Turkmenistan, Turkey, and Ukraine;

- Region 2: The Americas, including Greenland and the East Pacific;
- Region 3: The Far East, Australia, New Zealand, and the West Pacific.

2.3.1   User Spectrum

Although HTS terminals may operate in any of the satellite bands, there is a convergence toward HTS terminals operating in the Ka-band. Figures 2.4 and 2.5 provide charts summarizing the ITU frequency allocations as given in [2] for the Ka-band spectrum usually associated with the downlink and uplink, respectively. The notes in [2] are omitted from the charts for brevity.

In general, the frequency allocations are made on a per-region basis. Thus, looking at the downlink frequencies, the use of the band 17.3–18.1 GHz by geostationary-satellite systems in the fixed-satellite service (Earth-to-space) is limited to feeder links for the broadcasting-satellite service. In region 1, however, it can also be used for downlinks (space-to-Earth) but without claiming protection. (Note 5.516A of [2].) Downlinks (space-to-Earth) in region 2 are limited to broadcasting and not available in region 3.

Again, there slight discrepancies between the allocations of the three regions for the band 17.7–18.1 GHz and slight variations in different segments of the spectrum. The same applies for the bands 18.6–18.8 GHz and 19.7–20.1 GHz. The allocations for segments of the bands 18.1–18.6 GHz, 18.8–19.7 GHz, and 20.1–21.2 GHz are each identical for the three regions.

Examining the uplink frequencies from the ITU context, the band 27.0–27.5 GHz for satellite uplinks is limited to regions 2 and 3. The segments

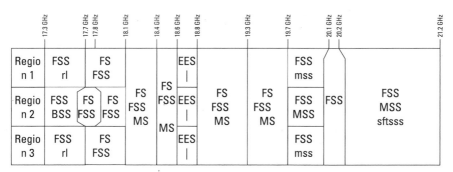

Figure 2.4   The ITU frequency allocations for the Ka-band (downlink).

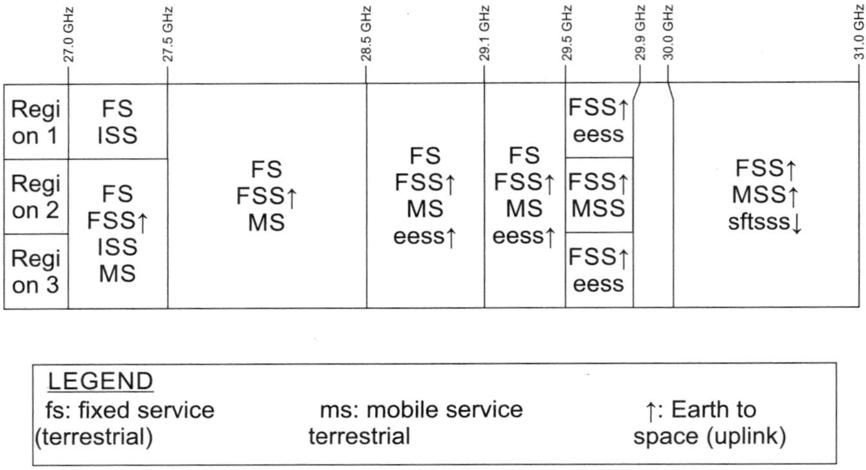

Figure 2.5 The ITU frequency allocations for the Ka-band (uplink).

of the bands 27.5 GHz–29.5 GHz and 29.9–31.0 GHz have identical allocations for the three regions, although there are variations between segments. The band 29.5–29.9 GHz has slight differences between the three regions. Table 2.1 provides the total spectrum available at the ITU level in the Ka-band for the downlink and the uplink.

It should be noted that although there is no such reference in [2], the top 1 GHz of these bands (i.e., the downlink band 20.2–21.2 GHz and the uplink band 30.0–31.0 GHz) are normally reserved for military/government usage.

Additionally, Note 5.516B [2] states that parts of the Ka-band spectrum (as identified in Table 2.2) are designated for use by applications in the high-density fixed-satellite service (HDFSS). This service is aimed at global broadband communication services throughout the world; the guidelines for their implementation are described in Resolution 143 [3]. HDFSS is a service delivered by HTS systems.

Table 2.1
Total Available Spectrum in the Ka-Band as Given in [2]

Region	Link	Lower Limit	Upper Limit	Total Spectrum
1, 2, and 3	Downlink (space-to-Earth)	17.3 GHz	21.2 GHz	3.9 GHz
1	Uplink (Earth-to-space)	27.5 GHz	31.0 GHz	3.5 GHz
2 and 3	Uplink (Earth-to-space)	27.0 GHz	31.0 GHz	4.0 GHz

Table 2.2
Parts of the Ka-Band Spectrum for Use by Applications in the HDFSS as Designated by Note 5.516B [2]

Downlink (Space-to-Earth)	Uplink (Earth-to-Space)
17.3–17.7 GHz in region 1	27.5–27.82 GHz in region 1
18.3–19.3 GHz in region 2	28.35–28.45 GHz in region 2
19.7–20.2 GHz in all regions	28.45–28.94 GHz in all regions
	28.94–29.1 GHz in regions 2 and 3

At the regional level, a number of regional bodies are responsible for the coordination of frequency use and spectrum requirements. These organizations, which have been set up in a cooperative arrangement between telecommunication administrations, are listed as follows:

- The European Conference for Posts and Telecommunications or the *Conférence européenne des administrations des postes et des telecommunications* (CEPT) for Europe;
- The Asia-Pacific Telecommunity (APT) for the Asia-Pacific region;
- The Comisión Interamericana de Telecomunicaciones (CITEL) for the Americas region;
- The African Telecommunications Union (ATU) for Africa;
- The Regional Commonwealth in the field of Communication (RCC) for Eastern Europe;
- The Arab Spectrum Management Group (ASMG) for the Arab countries in the Middle East and North Africa.

Joint proposals are submitted to ITU conferences by these regional bodies. These submissions are first agreed upon at the regional level. These bodies can also exercise a joint influence on policies to promote their regional interest, which is often the result of strong commercial motivation.

The CEPT represents 48 European countries. In the decision ECC/DEC/ (06)/03 [4], the CEPT [through the Electronic Communications Committee (ECC)] determined that terminals receiving in the band 19.7–20.2 GHz and operating below 60 dBW in the band 29.5–30.0 GHz are exempt from individual licensing requirements. This is in response to the need for harmonization of licensing regimes in order to control the installation, ownership, and use of radio equipment and to facilitate the provision of Pan-European services.

Additionally, there is a strong desire within the CEPT administrations to reduce the control exercised by administrations. This leads to a common EU regulatory framework for electronic communications. (The 19.7–20.2-GHz and 29.5–30.0-GHz bands are commonly referred to as the exclusive bands.)

Of course, decisions of the CEPT are not binding to the individual member countries. Each country remains sovereign. For ECC/DEC/(06)/03 [4] at the time of writing, 33 countries have implemented the decision; France has partially implemented the decision; three countries will not implement it; and the response of 11 countries is still pending. Out of the EU countries, 24 have implemented the decision, with one having only partly implemented it.

Similarly, a CEPT decision, ECC/DEC/(05)/08 [5], deals with the 17.3–17.7-GHz band. Thirty-two European countries have adopted the decision, three countries have not accepted the decision, and the decision of 13 countries is still pending. The situation in the EU regarding this decision is that 24 countries have agreed, while Belgium will not adopt the decision.

The inhomogeneity of the regulatory issues presents challenges, since designing for spectrum allocation on a country basis is difficult and could lead to system inefficiencies.

2.3.2   Gateway Spectrum

The discussion thus far has concentrated on the user spectrum. However, there is an equal capacity requirement for the gateways. The first generation of the Ka-band HTS (e.g., KA-SAT and ViaSat 1) employed the Ka-band for both the user links and the gateway links. Two solutions have emerged, and the two systems are examples.

One solution is to put the gateways in an area that is not included in the user service area. This allows frequency reuse to be fully utilized between the user beams and the gateway beams. This implies that the full spectrum is available for the user beams and that the same full spectrum is available for the gateways, but the two areas cannot overlap geographically. Orthogonality between the user beams and the gateways is achieved through spatial segmentation. The gateways must be outside the user service area. This is the approach adopted by ViaSat 1 where the user service area is comprised of the East Coast and the West Coast of the United States while the area in between is used to accommodate the gateways.

Another solution caters to a situation in which the gateways are embedded in the user service area. In this case, the spectrum has to be shared between the user service area and the gateways. Orthogonality between the user beams and the gateways is achieved through spectrum segmentation. Because in this case the spectrum is shared, the system capacity that can be achieved is lower, but it meets the operational and/or commercial requirement that the service area is not segmented. This is the approach adopted by KA-SAT where the user service area is a contiguous area covering Europe (pan-European) and parts of the Middle East with the gateways being positioned within the same area.

There is yet another solution: to use a different band for the gateways. The terminals are a well-established mass product, and therefore there is little motivation to change their band. Looking at the spectrum available for satellites that could be used for the gateways, there are the Q/V bands and the W-band.

Figure 2.6 summarizes the section of [2] relevant to the Q/V bands. The spectrum normally used for downlink is split in three segments (i.e., 42.5–43.5 GHz, 47.2–50.2 GHz, and 50.4–51.4 GHz). Figure 2.6 shows the allocations that can be used for the downlink with a thicker outline.

The first and third allocations, 42.5–43.5 GHz and 50.4–51.4 GHz, are applicable to all three regions.

The second spectrum segment, 47.2–50.2 GHz, actually includes six contiguous allocations for region 1 and four contiguous allocations for regions 2 and 3. Two allocations, namely 47.2–47.5 GHz and 47.9–48.2 GHz are applicable to all three regions. The remaining segments (i.e., 47.5–47.9 GHz, 48.20–48.54 GHz, 48.54–49.44 GHz, and 49.44–50.2 GHz) are similar in their allocations with the main difference being that in region 1 some segments support both uplink and downlink applications.

The uplink spectrum, 37.5–42.5 GHz, is structured in seven contiguous allocations. Six are common to the three regions. The 40.50–41.0-GHz allocation has slight differences between the regions. Table 2.3 summarizes the total available spectrum for the Q/V band.

Thus, 5 GHz of spectrum is potentially available for both the uplink and the downlink of the gateways. However, from a HTS payload design point of view, it is unfortunate that the uplink is segmented in three parts since this makes the payload more complex, necessitating extra equipment. Being part of the forward link, this is the most important link of the HTS system. It is also unfortunate that the uplink and downlink are adjacent at 42.5 GHz. This means that the full spectrum of either the uplink band and/or the downlink cannot be simultaneously exploited as a guard band is required. Since the forward link is the most critical in a HTS system, the downlink (i.e., the return link) is likely to be compromised. Section 2.4 further discusses the forward and return links.

Figure 2.7 illustrates the simple ITU allocation structure of the W-band. All allocations are common to the three regions. The downlink spectrum lies between 71.0 and 76.0 GHz with two adjacent allocations, 71.0–74.0 GHz

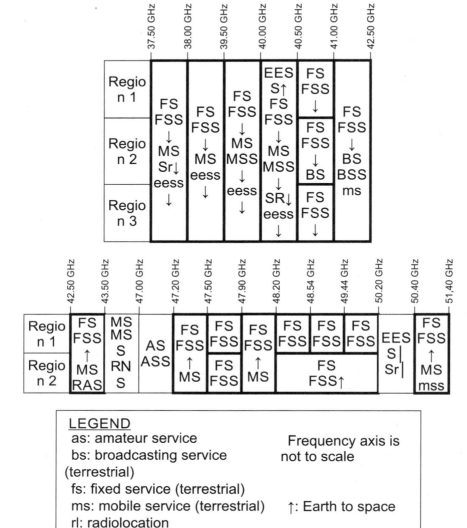

Figure 2.6 The ITU frequency allocations for the Q/V-bands (uplink and downlink).

Table 2.3
Total Available Spectrum in the Q/V-Band as Given in [2]

Region	Link	Lower Limit	Upper Limit	Total Spectrum
1, 2, and 3	Uplink	42.5 GHz	43.5 GHz	5.0 GHz
		47.2 GHz	50.2 GHz	
		50.4 GHz	51.4 GHz	
1, 2, and 3	Downlink	37.5 GHz	42.5 GHz	5.0 GHz

and 74.0–76 GHz. Similarly, the uplink spectrum spans between 81.0 and 86.0 GHz also with two adjacent allocations, 81.0–84.0 GHz and 84.0–86.0 GHz. Table 2.4 summarizes the total available spectrum for the W-band.

Similar to the Q/V band, 5 GHz of spectrum is available for the uplink and the downlink. However, each 5 GHz of spectrum is contiguous, and there is 5 GHz of guard band between the uplink and downlink. This would simplify the payload architecture, although equipment in this band is not commercially available. The challenge with achieving the required link availability remains, and it is unlikely that these bands will become usable on a worldwide basis.

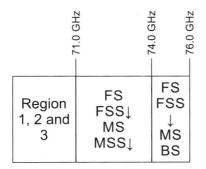 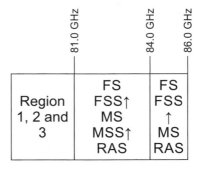

Figure 2.7   The ITU frequency allocations for the W-band (uplink and downlink).

Table 2.4
Total Available Spectrum in the W-Band as Given in [2]

Region	Link	Lower Limit	Upper Limit	Total Spectrum
1, 2, and 3	Uplink	81.0 GHz	86.0 GHz	5.0 GHz
1, 2, and 3	Downlink	71.0 GHz	76.0 GHz	5.0 GHz

2.4 Capacity

The *size* of the HTS system can be summarized as a single figure: the capacity in gigabits per second. Commercially, this is, of course, an important parameter, but caution is required in its use. There are many assumptions employed to arrive at this figure.

Very often this is the sum of the forward and return capacity. Technically, the return capacity may be easier to achieve, but the forward capacity is the system-dimensioning parameter and consequently more valuable commercially. Having the right mix is thus a design objective. Typically, the required return capacity is equal to 40% of the forward capacity.

The terminal parameters play a significant role. A given HTS system with the typical (70-cm) residential terminal may yield twice the forward capacity compared to the same system employing commercial aeronautical terminals. This is because the physical constraints and requirements in aviation also limit the radio frequency (RF) performance.

The system capacity is often quoted for clear sky conditions. It can be argued that for a large service area, the portion of the area affected by precipitation at any given time is small and that the portion of time when this happens is short, so that the net effect on the system capacity is small.

HTS systems have continued to use the link budget availability as the key parameter for the service-level availability (SLA). This is adequate for very high availability. For example, an availability of 99.8% implies an outage of 17.5 hours in a year. As HTS tends to target the consumers, the trade-off between availability and service delivery cost becomes important.

The most used approach to dimension the propagation margins of a satellite system is to use the methodology advocated by ITU-R in Rec ITU-R P.618-12 [6] and its dependencies. The general workflow is to get the probability distribution of the attenuation caused by the different atmospheric components—namely rain, cloud, gases, and scintillation—and to combine them in order to obtain the total attenuation distribution. From this distribution, the probability to be below a given margin and thus be available can be derived. For this analysis, the derived margins pertain to a long-term yearly average.

The main reason is that the meteorological parameters used as the inputs of the models to get the attenuation parameters are derived as yearly averages. Some models including monthly variability have been recently added to the margin computation workflow as ITU-R Rec P.840-8 [7] for cloud liquid water content and ITU-R Rec P.836-5 [8] for water vapor attenuation. In 2017, a new version of Rec. ITU-R P.837-7 [9] was adopted; this version supplies monthly maps of rain rate to predict monthly statistics of rain attenuation.

When the link availability is below the typical values, it is useful to characterize the outage in more detail, and the following questions become pertinent:

- What are the statistics of outages?
- What is the expected duration of an outage in a given month?
- When is an outage likely to happen in the day?

For example, the climate of South East Asia is affected by the Intertropical Convergence Zone (ITCZ), which introduces a large variability to the rainfall pattern depending on the time of the day and the season. A large portion of the precipitation falls in the late afternoon/evening over land, and the inverse happens for coastal/maritime areas. In a study by Onera [10], a methodology derived from Rec. ITU-R P.837-7 [9] and P.618-12 [6] has provided attenuation statistics on an hourly basis for the average day of a given month and for the number of outages for a given duration and depth.

2.5   Effective Isotropic Radiated Power (EIRP) and Receive Figure of Merit

The main RF parameters of a satellite are defined by the EIRP and the receive figure of merit [or gain-to-noise temperature ratio (G/T)]. In a classical communication satellite, each beam produces a coverage that is contoured and zoned with graded requirements for the EIRP and G/T.

In HTS systems, it is usual to specify the minimum level that corresponds to the service type to be delivered. Figure 2.8 illustrates the minimum EIRP requirement for Section AA' of the user service area as discussed in Section 2.4. Section AA' cuts a cross-section of the service area across a number of beams as shown in the service area part of Figure 2.8. Notice that the minimum requirements are typically met at the edge-of-coverage (eoc), which is also the interbeam junction. Depending on the satellite antenna technology, the difference between the eoc and the peak ranges between 3 and 6 dB. In the Ka-band, the uplink frequency band is roughly 50% higher in frequency than that of the downlink. For transmit/receive antenna technologies, the physical dimensions for the antenna system are the same for both the uplink and the downlink bands. However, the electrical dimension of the antenna is 50% larger at the uplink band. The center-to-center spacing between the beams is, of course, the same. This results in a higher peak-to-eoc difference for the uplink than the downlink.

Ideally, HTS systems employ a large multiplex per beam. Often, however, a number of multiplexes are employed (at least today) that do not necessarily use the same bandwidth. The EIRP may be specified in EIRP spectral density (dBW/MHz). Assuming that the EIRP density is constant for all multiplexes sharing the TWTA, this may simplify the link budget calculation per multiplex estimation, as illustrated in Table 2.5.

Recall that the capacity of HTS systems is typically quoted under clear sky conditions so that the first requirement would be the minimum required EIRP and G/T. However, the service availability is part of the offered quality of service, which means that an extra margin is required to ensure that the availability can be supported; see Figure 2.8.

It should be noted that in Figure 2.8, the system margin is shown as constant for the whole section. The system margin includes the propagation margin, which is a function of the local climatic characteristics and not, of course, constant over a large geographical section.

Most HTS systems employ an adaptive coding and modulation scheme (ACM). This means that the coding and modulation for each link is selected as function of the link quality. If the carrier-to-nose ratio (C/N) is high, then a high-order modulation is used, and the coding could be reduced. Conversely, if the C/N is low, the order of the modulation is reduced, and the coding could be increased.

The best-known ACM scheme for the forward link is given in the DVB-S2X standard [11], although other proprietary schemes exist. DVB-S2X is the successor of DVB-S2 that has been widely adopted in TV broadcasting. The improvements [12] of DVB-S2X with respect to DVB-S2 are listed as follows [13]:

- Low roll-off going down to 5%, allowing for smaller carrier spacing and advanced filter technologies;
- MODCOD and FEC upgrades (more granularity; adding 64, 128, and 256 APSK; improving FECs and MODCODs; and differentiating linear and nonlinear MODCODs);
- Wideband implementations, which are particularly useful in HTS;
- Very low SNR MODCODs to support mobile (land, sea, air) applications going down to E_s/N_0 of -10 dB.

Table 2.5
A Simplified forward Link Budget to Estimate the Capacity of a Beam

Forward Uplink		
Uplink EIRP	66.8 dBW	
Range	37,236 km	
Free-space loss	212.8 dBW	
Spreading loss	162.4 dB	
Satellite G/T	20.0 dB/K	
Input flux density	−95.7 dBW/m²	
C/N_0	102.5 dBHz	
Channel bandwidth	1,000 MHz	
C/N		12.5 dB
Forward Downlink		
Satellite EIRP	63.9 dBW	
EIRP density		33.9 dBW/MHz
Range	37,236 km	
Free-space loss	209.8 dB	
Earth station G/T	17.2 dB/K	
C/N_0	99.9 dBHz	
C/N		9.9 dB
Total		
C/N_0	98.0 dBHz	
C/N	8.0 dB	8.0 dB
Corresponding ideal MODCOD	16APSK	3/5-L
Spectral efficiency	2.4	
Roll-off	5%	
Capacity	2.29 Gbps	

Table 2.6 illustrates the operation of the ACM scheme based on the DVB-S2x standard. A beam with a peak-to-eoc difference of 4 dB is assumed. The data is provided for five contours namely: −4 dB, −3 dB, −2 dB, −1 dB, and 0 dB. Three sets of data are provided for clear sky conditions and for

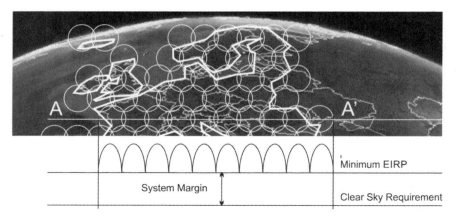

Figure 2.8 Demonstration of the minimum EIRP requirement for Section AA' of the service area depicted in Figure 2.2.

Table 2.6

MODCOD Table from DVB-S2x [11] (Table 20b: E_s/N_0 Performance at Quasi Error Free $FER = 10^{-5}$ (AWGN Channel) Normal FECFRAMES, 50 Iterations

Canonical MODCOD Name	Spectral Efficiency [bit/symbol][1]	Ideal E_s/N_0 [dB] for (AWGN Linear Channel)[2]	Ideal $C_{sat}/(N_0 \cdot R_s)$ [dB] (Nonlinear Hard Limiter Channel) (Informative)[3]
QPSK 2/9	0.434841	−2.85[4]	−2.45
QPSK 13/45	0.567805	−2.03	−1.60
QPSK 9/20	0.889135	0.22	0.69
QPSK 11/20	1.088581	1.45	1.97
8APSK 5/9-L	1.647211	4.73	5.95
8APSK 26/45-L	1.713601	5.13	6.35
8PSK 23/36	1.896173	6.12	6.96
8PSK 25/36	2.062148	7.02	7.93
8PSK 13/18	2.145136	7.49	8.42
16APSK 1/2-L	1.972253	5.97	8.4
16APSK 8/15-L	2.104850	6.55	9.0
16APSK 5/9-L	2.193247	6.84	9.35
16APSK 26/45	2.281645	7.51	9.17
16APSK 3/5	2.370043	7.80	9.38
16APSK 3/5-L	2.370043	7.41	9.94
16APSK 28/45	2.458441	8.10	9.76

Table 2.6 *(Continued)*

Canonical MODCOD Name	Spectral Efficiency [bit/symbol][1]	Ideal E_s/N_0 [dB] for (AWGN Linear Channel)[2]	Ideal $C_{sat}/(N_0 \cdot R_s)$ [dB] (Nonlinear Hard Limiter Channel) (Informative)[3]
16APSK 23/36	2.524739	8.38	10.04
16APSK 2/3-L	2.635236	8.43	11.06
16APSK 25/36	2.745734	9.27	11.04
16APSK 13/18	2.856231	9.71	11.52
16APSK 7/9	3.077225	10.65	12.50
16APSK 77/90	3.386618	11.99	14.00
32APSK 2/3-L	3.289502	11.10	13.81
32APSK 32/45	3.510192	11.75	14.50
32APSK 11/15	3.620536	12.17	14.91
32APSK 7/9	3.841226	13.05	15.84
64APSK 32/45-L	4.206428	13.98	17.7
64APSK 11/15	4.338659	14.81	17.97
64APSK 7/9	4.603122	15.47	19.10
64APSK 4/5	4.735354	15.87	19.54
64APSK 5/6	4.933701	16.55	20.44
128APSK 3/4	5.163248	17.73	21.43
128APSK 7/9	5.355556	18.53	22.21
256APSK 29/45-L	5.065690	16.98	21.6
256APSK 2/3-L	5.241514	17.24	21.89
256APSK 31/45-L	5.417338	18.10	22.9
256APSK 32/45	5.593162	18.59	22.91
256APSK 11/15-L	5.768987	18.84	23.80
256APSK 3/4	5.900855	19.57	24.02

[1] Spectral efficiencies are calculated in a bandwidth equal to the symbol rate R_s in case of no pilots. The corresponding spectral efficiency for a bandwidth equal to $R_s (1 + \text{roll-off})$ can be computed dividing the numbers in column "spectral efficiency" by $(1 + \text{roll-off})$.

[2] E_s is the average energy per transmitted symbol; N_0 is the noise power spectral density.

[3] C_{sat} is the hard-limiter pure carrier saturated power; $N_0 \cdot R_s$ is the noise power integrated over a bandwidth equal to the symbol rate. Performance results are for an optimized input back-off (IBO) and for a roll-off = 10%. $C_{sat}/(N_0 \cdot R_s)$ is equal to $E_{s,sat}/N_0$ and the difference between the E_s/N_0 of the AWGN linear channel and $E_{s,sat}/N_0$ is due to the compromise between operating back-off and nonlinear distortion (which is dependent on the roll-off).

[4] The FECFRAME length is 61 560.

5-dB and 19-dB rain fades. For illustration purposes, the rain fade is assumed to be constant within the beam. This is a mere simplification and does not reflect reality. The appropriate MODCOD is selected for each contour and corresponding rain fade so that the link performance is optimized for the prevailing conditions.

Looking at the clear-sky conditions, notice that the extra margin above the eoc is put to good use. In fact, the link is optimized for each site according the available link performance depending on the satellite performance at that location. This maximizes the system capacity and can be referred to as a static optimization. Looking at rain fade data, it can be seen that the link uses the MODCOD table to adapt to the prevailing conditions, which means that two sites on the same performance contour may operate at different MODCODs, because the rain fades are different and the system will continue to deliver the best with the available parameters. This can be referred to as dynamic optimization. The example in Table 2.7 shows that the system also has limitations. With a 10-dB rain fade at the eoc, the system runs out of MODCODs, and the link cannot be maintained. This is a simple example, and the climatic conditions tend to be localized and specific to the location. If a given availability is to be offered at service level, there will be a variance in the corresponding rain fade to the eoc contour of the system coverage so that the probability will be limited to certain areas. This shows the delicate balance between the dimensioning of the system and the offered availability. Thus, increasing the EIRP of such a HTS system by 1 dB would ensure that the system can offer the availability anywhere, but 1 dB corresponds to a 26% increase in satellite power, which comes at a cost.

Return link schemes also need to cater to a varying range of conditions. The complementary to DVB-S2X for the return link is DVB-RCS2 [13]. It operates in MF TDMA with carriers operating in QPSK, 8PSK, and 16PSK with a number of codes (1/2, 2/3, 4/5, 6/7). This allows terminals to uplink the required data using a combination of bandwidth, burst duration, modulation, and coding to allow terminals to make the best use of the available RF power under the link conditions.

2.6 Channelization

For HTS applications, maximum flexibility in managing the bandwidth is essential, and this is why most systems aim at maximizing the channel bandwidth. In classical payloads, the bandwidth tends to be limited to the RF

Table 2.7
An Illustration of the Operation of a MODCOD Scheme for a Beam with Contours at −4 dB, −3 dB, −2 dB, −1 dB, and 0 dB Under Clear-Sky Conditions and 5 dB and 10 dB

Canonical MODCOD Name	eoc → peak 1 dB Steps Clear Sky	eoc → peak 1 dB Steps 5-dB Rain Fade	eoc → peak 1 dB Steps 10-dB Rain Fade
QPSK 2/9			
QPSK 13/45			► ► ►
QPSK 9/20			►
QPSK 11/20		► ► ►	
8APSK 5/9-L		►	
8APSK 26/45-L		►	
8PSK 23/36	►		
8PSK 25/36			
8PSK 13/18			
16APSK 1/2-L			
16APSK 8/15-L			
16APSK 5/9-L			
16APSK 26/45			
16APSK 3/5	►		
16APSK 3/5-L			
16APSK 28/45			
16APSK 23/36			
16APSK 2/3-L	►		
16APSK 25/36			
16APSK 13/18	►		
16APSK 7/9	►		

power available from TWTAs. Thus, for DTH applications, the transponder bandwidths used are typically 27 MHz, 33 MHz, or 36 MHz, depending on the bands used.

The smaller satellite antenna beams of HTS system together with the higher frequency provide a higher antenna gain, which is a major asset at the system level. This means that a given EIRP spectral density can be maintained over a larger bandwidth.

If we consider a Ku-band–broadcasting satellite with an eoc EIRP of 51 dBW and 33-MHz channels, the EIRP density is 35.8 dBW/MHz. Assuming a typical television receive-only (TVRO) RF terminal and the MODCOD performance as given in DVB-S2X, about 100 Mbps can be delivered by the transponder. If we now consider a Ka-band HTS with a beam with an EIRP of 61.8 dBW over 50% of the beam area and over a total bandwidth of 1,000 MHz, the EIRP density is 33.8 dBW/MHz. This is 2 dB less than the Ku-band broadcasting case, but assuming a typical HTS terminal and again the MODCOD performance as given in DVB-S2X, over 2 Gbps of capacity can be delivered.

As usual, system analysis and comparison are complex since there are several parameters in a given system. Although the two systems are real, they employ different bands; however, the bandwidth is a significant performance enhancer. If we limit the HTS bandwidth to 33 MHz, the capacity will only increase to 160 Mbps. In both cases, a single carrier is assumed, which may be a bit optimistic for an HTS with 1,000 MHz today—but not necessarily tomorrow!

Both systems employ similar dimensions for their respective terminals. What is interesting is that in both cases, the assumed TWTA RF power is 130W, and therefore, the required DC power per resource is similar and the spacecraft antenna is about 2.5m in both cases. This means that the capacity delivered per unit payload mass and/or per unit spacecraft power is higher with a HTS system.

References

[1]   Fenech, H., et al., "VHTS Systems: Requirements and Evolution," *11th European Conference on Antennas and Propagation (EuCAP)*, Davos, Switzerland, March 20–24, 2017.

[2]   Article 5, *Frequency Allocations*; Chapter 2, Frequencies; Volume 1, Articles; Radio Regulations, Edition of 2016, International Telecommunication Union, Geneva, Switzerland, ISBN-10 : 9261199976, ISBN-13 : 978-9261199975.

[3]   *RESOLUTION 143 (Rev.WRC-07), Guidelines for the implementation of high-density applications in the fixed-satellite service in frequency bands identified for these applications*, Volume 3, Resolutions and Recommendations, Radio Regulations, Edition of 2016, International Telecommunication Union, Geneva, Switzerland, ISBN-10 : 9261199976, ISBN-13 :978-9261199975.

[4]   *ECC Decision (06)03*, Electronic Communications Committee, CEPT, approved March 24, 2006, amended March 8, 2019.

[5]   *ECC Decision (05)08*, Electronic Communications Committee, CEPT, approved June 24, 2005, amended March 8, 2013.

[6]   Recommendation ITU-R P.618-12 (07/2015), *Propagation Data and Prediction Methods Required for the Design of Earth-Space Telecommunication Systems*, P Series Radiowave Propagation, International Telecommunication Union, Geneva, Switzerland.

[7]   Recommendation ITU-R P.840-8 (08/2019), *Attenuation due to Clouds and Fog*, P Series Radiowave Propagation, International Telecommunication Union, Geneva, Switzerland.

[8]   Recommendation ITU-R P.836-5 (09/2013), *Water Vapour: Surface Density and Total Columnar Content*, P Series Radiowave Propagation, International Telecommunication Union, Geneva, Switzerland.

[9]   Recommendation ITU-R P.837-7 (06/2017), *Characteristics of Precipitation for Propagation Modelling*, P Series Radiowave Propagation, International Telecommunication Union, Geneva, Switzerland.

[10]   Fenech, H., et al., "Propagation Challenges in VHTS," *23rd Ka and Broadband Communications Conference [and 35th AIAA International Communications Satellite Systems Conference (ICSSC)]*, Trieste, Italy, October, 16–19 2017.

[11]   *Digital Video Broadcasting (DVB), Second Generation Framing Structure, Channel Coding and Modulation Systems for Broadcasting, Interactive Services, News Gathering and Other Broadcasting Satellite Applications, Part 2: DVB-S2 Extensions (DVB-S2X)*, ETSI EN 302 307-1 V1.4.1 (2014-11), European Telecommunications Standards Institute (ETSI), Sophia Antipolis, France.

[12]   Willems, K., "DVB-S2X Demystified," White Paper, Business Insight on DVB-S2X, March 2014, https://www.newtec.eu/frontend/files/userfiles/files/Whitepaper%20DVB_S2X.pdf.

[13]   *Digital Video Broadcasting (DVB); Second Generation DVB Interactive Satellite System (DVB-RCS2), Part 2: Lower Layers for Satellite Standard*, ETSI 301 545-2 V1.2.1 (2014-04), European Telecommunications Standards Institute (ETSI), Sophia Antipolis, France.

3

Specific Communication Mission Requirements

3.1   Overview

HTS systems were originally intended for residential broadband access. As we get more accustomed to the availability of broadband, the demand for broadband increases. Whether we are sitting in our living room, in an aircraft at 35,000 feet in altitude, offshore at sea, or even in a battlefield, we now expect the same connectivity. The demand is being driven predominantly by short message services (SMS), email, streaming, and video sharing among passengers and family members at home through smartphones, tablets, and laptops.

This chapter looks at the implications of connectivity demands for aviation, particularly in-flight connectivity and entertainment (IFEC); and for the maritime industry, which needs to consider passenger and crew connectivity; fleet management; and governmental issues, which includes command, control, and communications (C3); intelligence, surveillance, and reconnaissance (ISR); welfare; and the internet of things (IoT).

3.2   Aeronautical Services

3.2.1   The Market

There has been an increase in demand for inflight connectivity solutions from jets and private aircraft. Airlines are also focusing on the ancillary revenue, either directly through service fees or indirectly as a differentiator where the service is *free* but is intended to appear as a plus with respect to competing companies. Accordingly, airlines are offering services through inflight connectivity via media and content distribution, advertisements, and other related services.

Figure 3.1 shows the number of passengers carried worldwide on domestic and international flights and the total over the period between 2009 and 2018 as given by the International Civil Aviation Organization (ICAO). ICAO is a United Nations' specialized agency, established in 1944 to manage the administration and governance of the Convention on International Civil Aviation (Chicago Convention).

ICAO defines an international flight as one in which an aircraft takes off and/or lands in terminals that are in the territory of a state, other than the state in which the air carrier has its principal place of business. Unfortunately, this is not necessarily indicative of the flight time. A flight between New York and San Francisco is just under seven hours and, by this definition, a domestic flight, since both terminals are in the United States, while a flight from Seville,

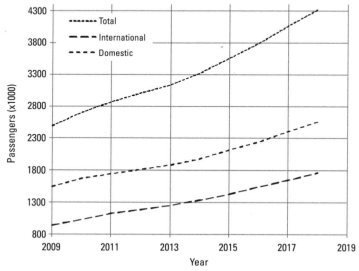

Figure 3.1   The number of air passengers carried over international and domestic flights and the total of the two segments in the period between 2009 and 2018 [1].

Table 3.1
The CAGR for the Number of Passengers Carried over International and Domestic Flights and the Total for the Two Sectors in the Period Between 2009 and 2018

International flights	7.3%
Domestic flights	5.7%
Total	6.3%

Spain, to Tangier, Morocco, which is under an hour, is an international flight. Nonetheless, most long-haul flights are international.

Table 3.1 shows the CAGR for international and domestic flights, along with the total for the two sectors. It can be observed that the CAGR for passengers flying on international flights is higher than that for those flying on domestic flights.

With the increase in travelers and people expecting connectivity anywhere they are—including on aircraft—and airline companies' efforts to enhance their clients' experience, the demand for IFEC is likely to continue to rise. In-flight connectivity (IFC) is a relatively recent service on satellites, with Boeing first introducing in-flight connectivity in 2004 as a commercial service through Connection by Boeing brand. Although related, in-flight entertainment (IFE) and IFC are not identical. IFE can employ cached technology where, for example, content like music and video is stored onboard without the need for connectivity outside the aircraft. Internet and phone service, on the other hand, require IFC. There are over 50 airline companies offering connectivity, but this represents perhaps only a small fraction of the 30,000-commercial aircraft in service.

3.2.2 The Service Area

In general, the aviation industry has exploited existing satellite capacity, allowing for a lower capacity cost and consequently a lower service cost and a faster time-to-market. This has led several IFC service providers to adopt the Ku-band, which remains a well-used band for this service. However, there are important areas that tend not to be as well served by satellite communications. These include the Atlantic and the Pacific Oceans.

The Atlantic is the busiest area, with over 1,737 transatlantic flights entering or leaving the EUROCONTROL network each day [2]. The bigger part of the traffic occurs over the North Atlantic between Europe and the United States. The traffic is characterized by a bi-directional flow: the eastbound flow and the westbound flow. Market requirements impose certain departure times. This implies that the westbound traffic flow crosses the 30°W longitude between 11:30 UTC and 19:00 UTC and that the eastbound traffic crosses the same longitude between 01:00 UTC and 08:00 UTC [3]. A simple model has been developed to show the west-east traffic profile over a 24-hour period; see Figure 3.2.

The model assumes symmetric traffic for the eastbound flow and the westbound flow. It assumes a truncated Gaussian distribution over a 24-hour period for each traffic direction, and the times given above are used to define sigma.

The bulk of the traffic in the early hours of the day is predominately the eastbound flow where the associated peak is seen to move eastward. The traffic for the middle hours of the day is dominated by the westward traffic.

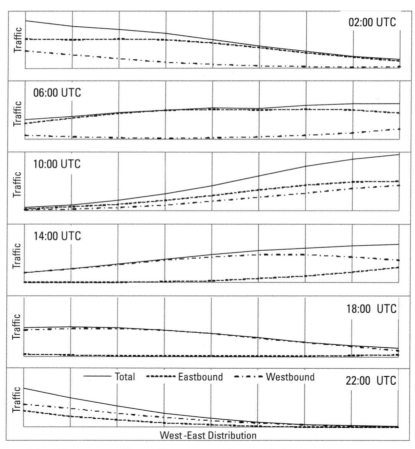

Figure 3.2   West-east traffic profiles over a 24-hour period on four-hour steps.

This is, of course, as expected, but what is interesting from a communication point of view is that it is the aggregate traffic that translates into the demand for communication capacity. The integral of profile is a measure of the total capacity demand, and it can be seen from the total traffic plots of Figure 3.2 that the first half of the day features higher demand than the second half.

Additionally, by looking along the total profile, it can be observed that the traffic over the west-east cut shows great variations at a given time and that for a given portion of the west-east cut there is also a great variance over a 24-hour period. These traffic variations reflect the capacity demand over the cut.

Of course, the communication system must be able to handle the total capacity, but it also needs to be made available where it is required. The traffic density in a cell can be a dimensioning factor. If, again, we consider the North Atlantic with track separation of a half-degree latitude and a longitudinal separation of 10 min [3] of flight, then a 1° cell could include less than 20 aircraft per flight level.

The situation can be very different around major airports. Hartsfield-Jackson Atlanta International Airport in the United States is the largest airport when considering the total number of passengers. Ranked by the number of international passengers, Dubai International Airport comes first, followed by Heathrow Airport in the United Kingdom.

Charles de Gaulle Airport, France, ranks sixth with 66 million international passengers and 422,000 aircraft movements in 2018. The longest runway of Charles de Gaulle Airport is 4,200m. During ground operations, an aircraft spends time at the apron boarding passengers, and engaging in pushback, taxiing, and takeoff. This is a relatively small area. If we consider the traffic to be integrated over the control traffic region (CTR) 1 Paris, the largest dimension of the area in the direction of the runways is about 30 NM. From a geostationary satellite point of view, this distance equates to 0.08°. This could be covered with a small beam but we need to consider about 200,000 passengers use the airport each day mostly concentrated over a few hours, resulting a high capacity density over a relatively short period.

At an airport, the peak capacity requirement occurs over the period when there is the maximum number of takeoffs and/or landings. The capacity for aeronautical IFC requirements over France is estimated and depicted in Figure 3.3 using a cell size of 0.1° as the resolution of the capacity requirement. The 100% maps by definition provide the peak capacity requirement per cell.

The main statistics of Figure 3.3 are given in Table 3.2. It will be noticed from the top row of Table 3.2 that the cell with the highest peak capacity requires no capacity for 90% of the time. This further illustrates the temporal variability of the capacity requirements [4].

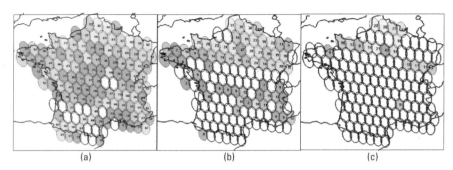

(a)   (b)   (c)

Figure 3.3 Maps showing the estimated capacity requirements for an IFC service over France using 0.1° cells for (a) 100%, (b) 95%, and (c) 90% of the time.

IFC traffic is characterized by the large variation in capacity requirement between cells and the large variations in capacity requirements over time for a given cell. This becomes a challenge when defining the capacity requirement of HTS systems where we would attempt to define capacity requirements per cell. A simple approach would be to identify the maximum capacity requirement per cell, but this could be technically inefficient and commercially expensive.

In this context, a sound approach is pooled resources, where, for example, instead of having power amplifiers per beam, a shared RF power resource is employed so that the available RF power is a given but the way it is apportioned over a number beams is flexible. One such technology is the multiport amplifier (MPA) where a set of amplifiers works as a single amplifier. As the name implies, the MPA is driven by a number of ports at the input and has an equal number of ports at the output. Thus the total RF power available over the total number of output ports is the total RF power available, but the apportionment of RF power per output port is an operational variable, set by the input drive corresponding to the given output port. Eutelsat 172 B employs MPA technology to provide an IFC system over the North Pacific in the Ku-band.

Table 3.2
The Capacity Requirements for the Percentage of the Time for the Cell with the Maximum Capacity for a Given Time Percentage

Maximum Capacity	100% of the Time	95% of the Time	90% of the Time
100%	224 Mbps	24 Mbps	0 Mbps
95%	211 Mbps	67 Mbps	47 Mbps
90%	211 Mbps	67 Mbps	47 Mbps

Generally, the apportionment of RF power is the most critical in such systems, but in order to maintain constant operating conditions for unaffected beam, it is important to maintain a constant EIRP density so as not to degrade the intrasystem interference level. This means that as the power is increased, the appropriate bandwidth has to be available to maintain the EIRP density. When possible, the maximum required bandwidth is provided to the cell, and with frequency reuse, this could be possible even though the frequency plan becomes more complicated. Digital processors can greatly facilitate bandwidth and RF power management.

3.2.3 The Terminal

A major challenge in IFC is the airborne antenna. By definition, this is mounted on the top side of the aircraft airframe with the objective of minimizing aerodynamic drag. This imposes mechanical constraints, particularly in terms of height, on the RF design of the antenna [5]. The following three antenna concepts can be envisaged:

- Antennas with mechanical steering in two axes, in azimuth and elevation. The aperture's largest dimension is often limited to 70 cm, and in order to limit aerodynamic drag, the height does not exceed 25 cm. The resulting aperture is asymmetrical.
- Phased arrays, which are intrinsically flat antennas and, therefore, interesting for their lower aerodynamic drag. They are electronically steered so there are no moving elements.
- Hybrid antennas that combine mechanical and electronic steering.

The aircraft beam of a flat antenna as seen at satellite is a function of the aircraft position. To the first order, the gain of a flat antenna is a function of the elevation. This is because the projected or effective aperture decreases as the elevation decreases with respect to the cosine of the scan angle, which is defined as zero at the antenna boresight. The pattern also changes with scan angle so that as the elevation decreases the beam also broadens in the elevation plan.

These flat antennas are typically planar arrays, and besides the cosine effect, there is also mutual coupling between the elements, which can further degrade the performance with scan angle. There could even be blind spots.

If for simplicity we assume a circular flat aperture with the aircraft flying level, the projected dimension of the antenna in the azimuth plane remains constant, but the effective dimension in the elevation is smaller. This is demonstrated in Figure 3.4. The effective area of the antenna and therefore the gain is, of course, reduced. However, the pattern is also modified. The dimension

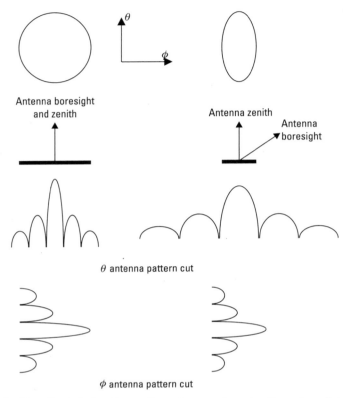

Figure 3.4 The effect of elevation on the antenna gain and patterns for a flat antenna.

in θ cut is unchanged so that the shape of antenna pattern is unchanged albeit lower due to the lower aperture area. In the φ direction the effective antenna dimension is smaller, and therefore the beam pattern is wider.

The effect on the communication link is dependent on the position of the aircraft relative to the satellite. At a low elevation to the north and south, the link budget suffers degradation in the signal level due to the reduced effective area and consequently antenna gain. The widening of the beam is in the elevation plane, which is perpendicular to the equatorial plane and therefore has no direct effect. At the same elevation in the east or the west, there is a similar loss in antenna gain but additionally, the widening of the beam that is still in the elevation plane, which is now aligned to the equatorial plane.

On the forward link (or downlink), this means that the link budget suffers a double hit: a loss of signal level due to the loss in antenna gain and an increase in adjacent satellite interference (ASI). On the return link (or uplink), regulatory effects limit the uplink off-axis EIRP to protect the adjacent satellites.

This means that the effect of the off-axis performance of the terminal is a function of the radial from the subsatellite point. This tends to limit the use of such antennas at low elevation (20° to 30°) especially in the east and the west.

Aircraft maneuvers contribute to the appropriate beam rotation. Pitch normally only occurs at the initial and final stages of the flight. Yawing introduces asymmetric flight and is normally avoided but could occur during turbulence. Rolling is employed during turns and could also occur during turbulence.

Gimbaled circular antennas do not suffer from this effect, as the frontal aperture is constant, and therefore antenna gain, and beam symmetry remain constant in all cases. However, the antenna height limitation often precludes a large circular antenna.

With a gimbaled rectangular antenna, the effective aperture is constant, so the antenna gain is independent of its geographical position. In an attempt to limit height, the radiation pattern in the elevation plane is sacrificed, which means that the ASI limits its operation in two symmetrical triangular regions centered on the equator to form a bowtie with the center at the subsatellite point.

The effect of varying performance according to the radial with respect to the subsatellite point is due to the skew angle, which is defined as 0° when flying on the satellite longitude and 90° when flying on the equator.

Figure 3.5 shows a range of aeroantennas. Figures 3.5(a, b) demonstrate the ThinKom solution, which employs separate transmit and receive antennas.

(a)   (b)

(c)   (d)

Figure 3.5   Photos of terminals used for IFC: (a, b) two terminals from ThinKom operating in the Ku-band and the Ka-band, respectively. (Photos courtesy of ThinKom Solutions.) (c) Two terminals from Viasat operating in both the Ku-band and Ka-band. (Photos courtesy of Viasat.)

Each antenna includes three circular plates that rotate together and differentially to control the azimuth, elevation, and polarization. The Viasat solution [Figure 3.5(c)] is based on a rectangular horn array with a mechanical control for elevation and azimuth. The antenna system is transmit-receive.

3.3 Maritime Services

3.3.1 The Market

The connectivity services in the maritime field are diverse and could cater to cruise ships, superyachts, cargo vessels, and research vessels.

There are approximately 45 cruise ships in operation of which the larger ones carry more than 4,000 passengers and a crew of 1,200 or more. With such a large concentration of people on a ship, passenger satisfaction and crew well-being become important. Cruise ships tend to operate in identified regions like the Caribbean and the Mediterranean Sea, but there are other operating areas including Alaska, the Baltic Sea, and the South Pacific. The Caribbean and Europe sectors represent 62.8% of the business, with 34.4% from the Caribbean and 17.3% from the Mediterranean. In 2019, the number of passengers was estimated to be 30 million. That means that nearly 20 million passengers are located in Europe (8.5 million) and the Caribbean (10.3 million). That eases the definition of the service area as the two well-defined geographical areas.

Ferry passengers are also a significant market. The busiest lines are across the English Channel, which alone carried almost 11 million passengers in 2019 [6]. The corresponding service area is also well-defined.

The Cruise Lines International Association Report of 2019 [7] names Instagram as the number one connectivity-centered trend in this business. Photos are driving interest in travel around the world. With on-board connectivity, cruise passengers are filling feeds with diverse travel experiences. This, of course, implies satellite connectivity in most cases. On-board smart tech is another identified trend to enhance the passenger experience and amongst the range of technologies involved, satellite connectivity must be one. Moreover, combining work with leisure time is on the rise. Straying far from the notion of device-free travel, many digital nomads are opting for trips where they can work remotely, reducing time off and lost wages. With satellite connectivity, passengers can keep up with work while enjoying a cruise vacation.

The term cargo ship is used to refer to a ship not carrying passengers. Cargo can include general cargo or packaged items; containers where goods are put into intermodal containers of standardized sizes; liquid cargo as carried in tankers; dry bulk cargo as coal and grain in loose form; and perishable goods like foods that need to be refrigerated. Ships for the latter are referred to as reefers.

According to the International Chamber of Shipping, there are over 95,402 merchant ships [8] trading internationally, transporting every kind of cargo. It is estimated that 1.5 billion seafarers are required to staff these vessels. Satellite communications can be used to boost the morale of the crew by ensuring that they stay in touch with family and friends and have access to social media, news, and entertainment even when operating in remote areas.

Slightly more than 50% of seafarers are officers, and the trend is that the portion of officers increases at the expense of ratings [9]. This is due to the fact there is a higher reliance on direct communication with their onshore offices. The amount of commercial radio telephone traffic handled at coast stations has diminished as a consequence.

There is also a larger demand for on-board telemetry and monitoring to be available in the onshore offices. Data generated from satellite information and on-board sensors, systems, and machinery are employed to support informed decision-making to optimize routes and asset management. Such technology can also be used to determine the most efficient route for the given meteorological and traffic conditions to estimate in real time the arrival time of vessels. Having more ship information allows for the identification of potential issues—so that ships' downtime can be reduced—and facilitates fleet management and maintenance. Intelligent containers with environmental sensors are used to monitor the conditions during transport.

Since cargo ships spend most of their time on the high seas where satellite communication is the only option, there is a tendency to rely more on satellite communications for support, management, maintenance, and security. The availability of satellite communication also supports the well-being of crew and can thus be a differentiator for attracting staff.

3.3.2 The Service Area

The service area requirements tend to depend on the service targeted. For cruise ships, the requirements are relatively contained with the biggest being the Caribbean and the Mediterranean. Most of the capacity contracts tend to be with cruise lines and satellite operators or service providers. Some cruise lines operate specifically in a given region; others operate in one region for part of the year and in another for the other part of the year.

To some extent, this simplifies the service area definition, since, for example, the Caribbean and the Mediterranean are well-defined regions that tend to be relatively well-covered by satellite communications. However, there could be an inventory issue. When ships are in one area for part of the year and in another for yet another part of the year, one solution could be to lease the required capacity in the two areas. Unfortunately, that is economically inefficient for the service provider. Another solution could be to lease capacity for part of the year in one region and in another region for the other part of the year. That would be inefficient for the satellite operator, since there could be unsellable capacity when the ships are absent.

Alternatively, satellite systems could be defined where the same capacity could be shifted geographically according to the requirements [10]. In fact, most large cruise ships carry a few thousand people, including passengers and crew, and there are a relatively small number of them (perhaps 100) in a given region. In addition, cruise ships are relatively slow-moving. In consideration of these factors, a system could be conceived where the service area could be dynamically optimized for the given number of points (or ships) to offer the best performance in the given situation. Larger cruise ships have pushed the capacity requirement per ship up by two orders of magnitude from a few megabits per second to hundreds of megabits per second.

The service area for cargo ships tends to be more challenging, because there are more of them; also, there is less crew (and, of course, no passengers), and the ships are spread over a vast area. Figure 3.6 illustrates the spread of cargo shipping activity in the world [11]. Although shipping routes cover practically all parts of the oceans, the northern hemisphere tends to be the busiest. Thus, global systems such as INMARSAT tend to offer a good global solution, especially for the remote areas of shipping. However, a lot of shipping follows the coast. Such areas include the Mediterranean, the Baltic, the English Channel, the Red Sea, the Gulf, the Malacca Strait, and the area around Taiwan and Japan, as can be seen in Figure 3.6. These regions are well covered by a regional HTS system.

3.3.3 Terminals

The capacity requirements on large cruise ships could be in excess of 100 Mbps, and consequently the required antenna aperture could be as large as 2.4m. These systems could be tri-band so that they would be operational in the C-band, Ku-band, and Ka-band and could be used with geostationary satellites or medium-Earth orbit satellite systems like O3B.

The requirements for cargo ships are considerably different from those for cruise vessels. The capacity could be an order of magnitude or two lower, but the space available may also be more limited, and often the

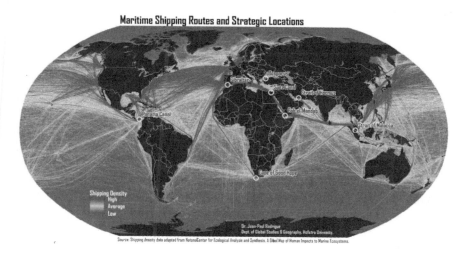

Figure 3.6   Cargo shipping routes over the world.

antenna aperture is smaller in the range of 60 cm to 1.5m according to the requirements.

Figure 3.7 illustrates an array of maritime terminals covering different applications. At the lower end, there is small terminal employing the Ka-band,

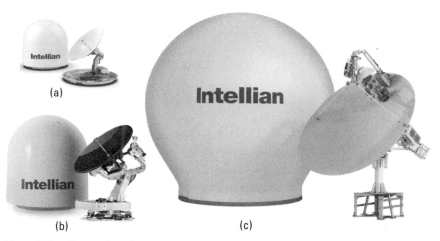

Figure 3.7   Photos showing a range of maritime terminals: (a) A 60-cm Ka-band terminal suitable for the larger pleasure craft and smaller ships, (b) a 1.5-m Ku-band terminal for cargo ships and ferries, and (c) a 2.4-m tri-band system for larger cruise ships. (Photos courtesy of Intellian Technologies, Inc.).

suitable for pleasure craft and smaller ships. At the high end, there is large terminal that can operate in the C-band, Ku-band, and Ka-band. This targets the larger cruise ships. There is also a moderately sized terminal as used on ferries and cargo ships operating in the Ku-band. These are mere examples, and the range is larger.

The physical volume requirement is less stringent in the maritime field, and often a classical parabolic reflector configuration is employed, although when the antenna has to operate in a number of bands the RF design becomes more demanding.

Ships do not intrinsically provide a stable platform. They will pitch, roll, and jaw according to the sea state and navigation. Thus, antennas need to be stabilized so that they are always pointing in a given direction independent of the ship's attitude. This, of course, adds considerable complexity to the antenna system. Moreover, salt water is quite an aggressive environment that is rather unfriendly to electronics and RF and mechanical equipment; as a result, radome is required.

In order to maintain seamless communications under all conditions, more than one antenna system is often used. This ensures that an unobstructed view is available from an antenna in all orientations of the ship and that there is no blockage by the super-structure of the ship. It can also be used to increase reliability.

3.4   Governmental Services

The demands and requirements of governmental and military users were a key driver in the early development of satellite communications systems, and today satellites remain an essential element of C3 and ISR. The communication requirements for governmental and military applications can be divided into three sectors: core, extended core, and augmentation.

Typically core capacity uses sovereign satellites that tend to be hardened to withstand environments in space that are harsher than the usual conditions in space. This implies that the equipment used aboard the satellite even for standard functions is potentially different in design with respect to commercial satellites and also uses different components. The testing is also more intensive than that of commercial satellites. As the name implies, core capacity tends to be expensive and limited.

There are applications where core capacity is not available either because of surge demand or because core capacity is not available in the area of interest.

The requirements may be identical to that of core capacity although commercial capacity is used instead. This is where extended capacity is used.

There are yet other applications where although the capacity is intended for governmental or military personnel, it is noncritical in nature. This is welfare capacity that is used by soldiers to keep in touch with the family, social media, news, and entertainment. This capacity is not core and tends to be carried on commercial satellites.

As communication requirements increase, there is a tendency toward the use of a balanced mix of commercial and military capacity. Furthermore, the balance is expected to tip further in the direction of commercial satellite systems due to pressures on national defense budgets, the increasing need for resilient satellite communications (SATCOM) architectures, the adoption of distributed systems to counter cyberattack, and the increasing need for global ISR [12].

The government services business already represents a significant portion of the revenues of the top three commercial satellite operators. Table 3.3 shows the revenue from the governmental services for Intelsat S.A, SES S.A, and the Eutelsat Communications Group.

It should be noted that although Table 3.3 quotes all figures for a year, the financial year does not have the same definition for the three operators. Nonetheless, the total for the three satellite operators adds up to 877 M$ (780 M€). It can be considered that most of this revenue comes from classical capacity so that it could be speculated that given HTS capacity in the required areas there is scope for further growth.

Military communications convey a mixture of voice and data (with growing applications to include telemetry, imagery, texting, file transfer, remote sensor computer access, paging, email and Internet, and facsimile) extending

Table 3.3
Revenue from the Government Services for Intelsat S.A, SES S.A, and the Eutelsat Communications Group

Satellite Operator	Revenue		Year ending
Intelsat S.A.	352.8 M€	392.0 M$ [13]	31 December 2018
SES S.A.	275.4 M€ [14]	306.0 M$	31 December 2018
Eutelsat Communications Group	161.5 M€ [15]	179.4 M$	30 June 2019
Total	789.7 M€	877.4 M$	

to teleconferencing and streaming video of varying levels of mission criticality. There is a trend to deploy fewer personnel but provide better support including better communications among the soldiers but also between the soldiers and the upper ranks and political leaders. This implies that more capacity is required locally in the field and back to the country.

Commercial satellite systems offer a logistical advantage as they are more prolific. For example, a remotely piloted aircraft system (RPAS)—also known as a drone or unmanned aerial vehicle (UAV)—might be required to be deployed anywhere in the world. The probability that commercial Ku-band capacity is available anywhere the RPAS is required to operate is higher than that of the sovereign capacity being available in that area.

The availability of HTS is a great attraction for military applications, especially considering the size of terminals, the delivered capacity, the discretion, and the cost of service acquisition and operational costs. Of course, it is possible for a country to have a sovereign HTS system, but it would probably be difficult to justify the cost considering that the cost of service is due to the leverage by the number of users. It can be considered that in general the cost per user decreases as the size of the system increases. HTS systems are designed for 100,000 users or more.

Since HTS systems are designed for a large population of users, redeployment is not a realistic option for two reasons. The services to the population from a given orbital location cannot be terminated because capacity is required elsewhere. Additionally, the gateways do not tend to be very portable since they are often served with dedicated satellite beams with a specific geographic mapping. Therefore, it would be more attractive to use regional systems on an ad hoc basis according to the requirements. For a sovereign capacity solution, several HTS systems would be required to cover all the regions of interest, which would imply high costs and low overall fill rates, thereby making the service cost even higher.

HTS systems deliver capacities to users that are at least an order of magnitude higher than those delivered by classical satellites, including military ones. In an environment where military applications are demanding more capacity and where conflicts are increasingly asymmetrical, not embracing HTS systems may actually push the balance in the wrong direction.

It is no longer adequate for a country to develop a new technology; it is now essential that that countries integrate and adapt a new technology in its military applications. Current military processes are not sufficiently responsive to need. Rather, they are over-optimized for exceptional performance at the expense of providing timely decisions, policies, and capabilities [16]. The cycle time from technology development to system application and service provision seems to be getting faster in the commercial world, and although this has not been traditionally the modus operandi in the military world, it could be become an opportunity for the service adoption and upgrade to be faster and to be attained at a lower cost while at the same time opening the geographical scope.

There is also a trend for dual-use Ka-band systems. The commercial Ka-band and the military Ka-band are actually adjacent. The satellite equipment needs to operate over a larger bandwidth. This can be managed, and there are payload designs that can be envisaged to offer the required operational flexibility. The impact on the ground equipment may be negligible, especially if the services are kept segmented. If the military terminal is only to operate in the military Ka-band, there is no impact on the military terminal. Of course, it may be attractive for the terminal to operate in both bands to maximize operational flexibility and agility.

HTS systems offer more resilience to interference or jamming. The fact that the coverage is composed of several cells implies that at worst a cell is jammed per jammer. Additionally, electromagnetic waves are not limited to cell definition. Especially toward the beam edge, it may be possible to use resources from a neighboring beam. The carrier-to-interference ratio (C/I) may degrade, but considering that the spectrum used by the adjacent cells and/or the polarization is different, service could be maintained [17]. Jamming the gateway is also technically possible but difficult. The gateways may not be in the same region and typically have a very narrow beam so as to maximize the performance. However, this also means that the jammer would be very close to within, for example, 100 km and similarly dimensioned.

3.5 IoT

In 2012, the ITU Recommendation ITU-T Y.2060 [18] defined IoT as a global infrastructure for the information society, enabling advanced services by interconnecting (physical and virtual) things based on existing and evolving interoperable information and communication technologies. It clarifies that through the exploitation of identification, data capture, processing, and communication capabilities, the IoT makes full use of things to offer services to all kinds of applications, while ensuring that security and privacy requirements are fulfilled. Additionally, it notes that from a broader perspective, the IoT can be perceived as a vision with technological and societal implications.

But what is a thing? It is an object in the physical world (physical things) or in the information world (virtual things). Things must be capable of being identified and integrated into communication networks.

The IoT, a relatively new service that has developed over the last decade, has given rise to a new wireless network technology: low-power wide area network (LPWAN). IoT applications include the following:

- Security (e.g., managing surveillance systems remotely);
- Asset tracking (e.g., geolocation of hazardous goods and monitoring their environmental conditions, including lock state and temperature);
- Smart agriculture where sensors measure soil parameters such as humidity and pH factor, so that water and fertilizers can be applied accordingly;
- Smart metering/smart grid systems, in which utility companies require real-time grid monitoring to make decisions on load sharing, outages, and interruptions;
- Smart homes where owners can remotely control or monitor their homes' temperature, humidity, media and electrical appliances, security, and water flow to prevent damage;
- Smart cities with security, safety, traffic monitoring, and management together with efficient management of resources like heating, electricity, water, and parking spaces;
- Manufacturing automation where real-time machinery monitoring can help maximize efficiency and productivity through remote control;
- Governmental use (e.g., ISR).

The requirements for IoT applications are quite diverse and can include capacities of a few bits per day, long battery autonomy, a high quality of service (QoS), higher capacity, and within or away-from digital cellular coverage.

It is, therefore, not surprising that there is no single solution for the spectrum of applications. There are solutions that employ Third-Generation Partnership Project (3GPP) standards and others that do not [19]. (The 3GPP is a standards organization that develops protocols for mobile telephony within the scope of the ITU's International Mobile Telecommunications-2000.)

The narrowband IoT (NB-IOT) [20], a 3GPP system that uses 200-kHz channels in the cellular licensed spectrum, can coexist with other cellular services. The specification was frozen in 3GPP Release 13 in June 2016. The data rate is limited to 200 kbps for the downlink and to 20 kbps for the uplink. The maximum payload size for each message is 1,600 bytes. Because it uses channels within a licensed spectrum, the QoS can be ensured, but this adds to the service cost.

The non-3GPP systems typically use unlicensed spectrum. In Europe, there are several bands allocated to IoT applications [21]. One of them, the band 862.0–874.4 MHz, limits the EIRP to typically 25 mW with a duty cycle of 0.1%, although there are several subbands with specific limits. This spectrum is often referred to as the 868-MHz band. Note 5.150 of the Radio Regulations [22] establishes the 902–928-MHz band as an industrial, scientific, and medical (ISM) band. This is commonly called the 915-MHz band.

Developed in 2010, the Sigfox system [19, 20] uses proprietary techniques. It uses the 868-MHz band in Europe, the 915-MHz band in North America, and the 433-MHz band in Asia. The uplink rate is 100 bps in 100-Hz channels. The message payload is limited to 12 bytes, and there is also a limit on 140 messages a day.

The LoRa [19, 20] system was developed in 2009 but was standardized as LoRaWAN in 2015. It also uses unlicensed bands and uses spread spectrum to provide a range-data rate trade-off. The data rate is between 300 bps and 50 kbps depending on the spreading factor and bandwidth. The payload length is limited to 243 bytes.

Where do HTS systems fit into the IOT? Figure 3.8 shows a possible implementation. Figure 3.8(a) illustrates the classical network where the devices communicate with the gateway in their native band in a bi-directional way. The gateway then passes the messages to the network server on the cloud.

Figure 3.8(b) shows the same topology as in the classical case except that instead of the direct connection between the gateway and the network server,

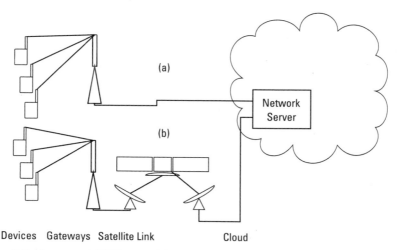

Figure 3.8 A terrestrial IOT network including a satellite link: (a) classical network with bi-directional communication and (b) a network with a satellite link.

a satellite link is included. The satellite link can, of course, employ any of the satellite bands. Considering the data being handled and the cost requirements, it is important that the satellite link is dimensioned for the requirements: performance and cost. Eutelsat's IOT FIRST is such an application where the gateway uses a small terminal with an aperture of 60–74 cm in the Ku-band, but it could also be an excellent application for a HTS system.

IoT services can also be directly delivered via satellite. Table 3.4 shows a link budget based on the Sigfox requirements. The object's EIRP is assumed to 14 dBm; this corresponds a RF power of 25 mW into an antenna with a 0–0 dBi gain. Only the return uplink is shown as it is assumed that the space-borne receiver has an equivalent performance to the gateway receiver so that the payload is regenerative. Of course, considering that the intention is to have compatibility with the standard IoT objects, an unlicensed band

Table 3.4
A Simple Link Budget for a Satellite-Based IOT System

	Beam Center	eoc
Uplink Frequency	865 MHz	
Uplink EIRP	14.0 dBm	
Satellite Altitude	550 km	
3-dB Full Beamwidth at Satellite	61.7°	
Antenna angle	0.0°	42.0°
Coverage on Earth's Surface	2,058 km	
Elevation	90.0°	43.4°
Slant Range	550 km	768 km
Free-space loss	146.0 dB	148.9 dB
Polarization Loss	3.0 dB	
Satellite Antenna Gain	5.1 dBi	−0.5 dBi
System Noise Temperature	27.0 dBK	
Satellite G/T	−21.9 dB/K	−27.5 dBHz
C/N_0	41.7 dBHz	33.2 dBHz
Signal Bandwidth	100 Hz	
E_s/N_0	21.7 dB	13.2 dB
Required E_s/N_0	8.0 dB	
Link Margin	13.7 dB	5.2 dB

is used. A LEO satellite at an altitude of 550 km is assumed. A simple three-turn helix is assumed for the payload uplink antenna for coverage that has a 2,000-km diameter on Earth.

Since there is a healthy margin over the coverage, it can be concluded that such a system is feasible. This is the basis of the ELO constellation of 25 LEO nanosatellites that Eutelsat has planned to be operational 2022 in partnership with Sigfox [23].

However, could the system work from GEO? Theoretically, the beam parameters would have to be maintained, which means that the beamwidth would have to be reduced from 61.7° to 1.2°. Reducing the beamwidth would increase the antenna gain by approximately 33 dB, which is equivalent to the increase in free-space loss. The spacecraft antenna would be challenging. If it were to be a helical antenna, it would have to be increased to more than 8,500 turns. If a reflector antenna were used, it would have to be about 16m. A stand-alone GEO solution may be technically possible, but it may not be compatible with the service economics.

References

[1] *Presentation of 2018 Air Transport Statistical Results*, International Civil Aviation Organization (ICAO), Montreal, Canada, https://www.icao.int/annual-report-2018 /Documents/Annual.Report.2018_Air%20Transport%20Statistics.pdf#search=The% 20World%20of%20Air%20Transport%20in%202018.

[2] "Celebrating 100 years of Transatlantic Flights," *Eurocontrol News*, June 17, 2019, https://www.eurocontrol.int/news/celebrating-100-years-transatlantic-flights.

[3] *North Atlantic Operations and Airspace Manual*, NAT Doc 007, V.2019-3 (Applicable from July 2019), prepared by the ICAO European and North Atlantic Office on behalf of the North Atlantic Systems Planning Group (NAT SPG).

[4] Fenech, H., et al., "Satellite Antennas and Digital Payloads for Future Communication Satellites," Special Issue on Recent Advances on Satellite Antennas for Communication, Navigation, and Scientific Mission Payloads, *IEEE Antennas and Propagation Magazine*, October 2019.

[5] *Satellite Earth Stations and Systems (SES); Technical Report on Antenna Performance Characterization for GSO Mobile Applications*, ETSI TR 103 233 V1.1.1 (2016-04), European Telecommunications Standards Institute (ETSI), Sophia Antipolis, France.

[6] U.K. Department for Transport, U.K. International Short Sea Passengers by Ferry Route, SPAS0102, Sea Passenger Statistics: Data Tables, August 2018, Last updated January 8, 2020, https://www.gov.uk/government/statistical-data-sets/sea-passenger -statistics-spas#all-uk-international-short-sea-long-sea-and-cruise-passengers.

[7]   *2019 Cruise Trends & Industry Outlook*, Cruise Lines International Association, Inc., 2019, https://cruising.org/news-and-research/-/media/CLIA/Research/CLIA-2019-State-of-the-Industry.pdf.

[8]   *Review of Maritime Transport 2019*, United Nations Conference on Trade and Development (UNCTAD), 2019, https://unctad.org/en/PublicationsLibrary/rmt2019_en.pdf.

[9]   *Global Supply and Demand for Seafarers*, International Chamber of Shipping, http://www.ics-shipping.org/shipping-facts/shipping-and-world-trade/global-supply-and-demand-for-seafarers.

[10]   Fenech, H., and S. Amos, "Eutelsat Quantum—Class Satellites Answering the Operator's Need for Flexibility," *3rd ESA Workshop on Advanced Flexible Telecom Payloads 2016 ESA/ESTEC*, Noordwijk, The Netherlands, March 22–25, 2016.

[11]   Rodrigue, J.-P., *Shipping Density Data Adapted from the National Center for Ecological Analysis and Synthesis, A Global Map of Human Impacts to Marine Ecosystems*, Dept. of Global Studies & Geography, Hofstra University, Hemstead, NY, 2020 https://transportgeography.org/contents/chapter5/maritime-transportation/domains-maritime-circulation/.

[12]   Fenech, H., et al., "How Commercial Satellites Satisfy Military Requirements," *23rd Ka and Broadband Communications Conference and 35th AIAA International Communications Satellite Systems Conference (ICSSC)*, Trieste, Italy, October 16–19, 2017.

[13]   *Intelsat Announces Fourth Quarter and Full-Year 2018 Results*, Intelsat S.A., http://www.intelsat.com/wp-content/uploads/2019/02/4Q18ERFINAL.pdf.

[14]   *Full Year 2018 Results*, Press Release, SES S.A., https://www.ses.com/press-release/full-year-2018-results.

[15]   *Consolidated Financial Statements as of June 30, 2019*, Eutelsat Communications Group, https://www.eutelsat.com/files/PDF/investors/EC_comptes%20consolid%c3%a9s_annuels_FY19_EN.pdf.

[16]   *Summary of the 2018 National Defense Strategy of the United States of America:* Sharpening the American Military's Competitive Edge, https://dod.defense.gov/Portals/1/Documents/pubs/2018-National-Defense-Strategy-Summary.pdf.

[17]   Dutronc, J., H. Fenech, and E. Lance, *Method for Establishing Radio Links by Means of a Multi-Beam Satellite, Patent*, Eurpoean Publication: EP2099142 (B1), granted 21 November 2012, French Publication: FR2928511 (B1), granted on December 17, 2010, US Publication: US8594661 (B2), granted November 26, 2013.

[18]   Recommendation ITU-T Y.2060, *Overview of the Internet of Things, Series Y: Global Information Infrastructure, Internet Protocol Aspects and Next-Generation Networks, Next Generation Networks—Frameworks and functional Architecture Models*, International Telecommunications Union, Geneva, Switzerland, approved June 15, 2012, published 2013.

[19]   Sami, T., "IoT Standards Part I: IoT Technology and Architecture," *Training on Planning Internet of Things (IoT) Networks*, ITU, Bandung, Indonesia September, 25–28, 2018.

[20]   Mekkia, K., et al., *A Comparative Study of LPWAN Technologies for Large-Scale IoT Deployment*, *ICT Express*, Vol. 5, No. 1, 2019, pp. 1–7.

[21]   ERC Recommendation 70-03, *Relating to the Use of Short-Range Devices (SRD)*, Electronic Communication Committee, CEPT, Tromsø 1997 with subsequent amendments June 7, 2019.

[22]·   Article 5, *Frequency Allocations; Frequencies*; Chapter 2, Volume 1, Articles; Radio Regulations, Edition of 2016, International Telecommunication Union, Geneva, Switzerland, ISBN-10 : 9261199976, ISBN-13 : 978-9261199975.

[23]   "Eutelsat Kicks Off ELO, Its Constellation of Nanosatellites Dedicated to the Internet of Things," press release, Eutelsat, September 24, 2019.

4

The Evolution of the Current HTS Payload Architectures

4.1 Overview

This chapter takes us from the classical payloads to the current HTS payloads. It looks at the functional features of the architecture of a classical payload and addresses its building blocks and their impact on operational requirements.

The HTS payload architecture has evolved to meet the specific functional requirements associated with broadband access. This chapter discusses the evolution to today's HTS payloads, which consist of two functional payloads (i.e., the forward and the return payloads). This structure has an impact on the organization of frequency plans for HTS systems. The chapter also compares classical payloads and HTS payloads, applying color schemes to the frequency plan with the ensuing frequency reuse and to the antenna system to produce a feasible system with improved antenna performance.

Since HTS payloads tend to be large, it is important to use equipment efficiently. The chapter addresses some of the optimization that makes it possible to reduce the equipment count while maintaining performance and, particularly, system capacity. It also examines associated specific challenges.

4.2   Classical Payloads

Figure 4.1 shows a functional block diagram of a classical payload, in which the uplink signal is received by the uplink antenna. The design of this antenna is closely associated to the uplink service area. The shape of the coverage could be circular or elliptical, but in most classical payload, the shape is determined by the commercial requirements and could be shaped to better represent the service area.

The signal from the uplink antenna is amplified by the low-noise amplifier (LNA). The performance of the LNA and the antenna heavily influence the figure of merit of payload receive performance, which is expressed as the G/T. The LNA performance in terms of gain and noise figure (or noise temperature) is constant over the service area, although there could be variations across the uplink band. Typically, the LNA amplifies the whole uplink spectrum so that it is a wideband unit. Variations in the G/T over the service area are mostly due to variations in the antenna gain.

The noise figure of LNAs is a function of their physical temperature. Accordingly, some payload designs have employed thermal subsystems for the LNA in order to lower the physical operating temperature to improve the LNA noise figure and ultimately the payload G/T. Such thermal subsystems can include a dedicated heat pipe system and a radiator system to efficiently radiate the heat generated by the LNA and maintain a low operating temperature. They could also be part of the spacecraft thermal system where a cold area is available by design. Typical physical operating temperatures for the LNAs when such thermal systems are employed are in the range of 0°C–20°C instead of 20°C–40°C.

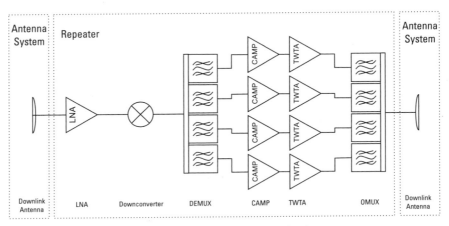

Figure 4.1   A functional block diagram of a classical payload.

The uplink spectrum of a given satellite band is normally at a higher frequency than that of the downlink spectrum; therefore, the spectrum must be downconverted. The uplink spectrum is often composed of a number of subbands, and it is not very likely that the translation frequency is the same for each subband. This is especially so, in the Ku-band where there could be six uplink subbands, which could necessitate up to six converters with each operating over a reduced bandwidth. Thus, converters are also wideband units although potentially not as wide as the LNA.

The converters feed into the demultiplexer (DEMUX), also known as the input multiplexer (IMUX). The DEMUX is the interface between the input wideband section of the payload and the channelized section. The DEMUX is a series of bandpass filters, one for each channel with a signal distribution system so that a frequency-multiplexed signal at the input ends at a number of ports with each channel available at a specific port. The distribution system is either a network of hybrids or a network of isolators.

Typical channel bandwidths at Ku-band and below are 27 MHz, 33 MHz, 36 MHz, 54 MHz, and 72 MHz. The filters are typically tenth-order elliptical filters that attempt to maintain a flat in-band performance and rejection outside the passband. Traditionally, these filters employed waveguide cavity technology, but most current designs use dielectric technology at least at the Ku-band and below.

The DEMUX feeds the amplification chain, which includes the channel amplifier and the high-power amplifier (HPA), which is often a TWTA. The channel amplifier (CAMP) has a signal-conditioning function that includes the low-level amplification to provide the drive power for the intended operating point of the HPA and the selection of the operating mode. In the payload-operating mode there are two main operating modes, described as follows:

- The *fixed-gain operating mode* (FGM for fixed gain mode) in which the CAMP is set to operate at a given set gain. This is best suited for uplinks from multisources where the operating condition must be guaranteed under a wide range of input conditions (e.g., a single uplink or the dimensioning case where the maximum number of uplinks occur). The situation may not be definitive as a statistical approach may be adopted. This means that an operating point of the HPA may be different for different percentages of the time or different numbers of simultaneous uplinks.
- The *automatic gain control* (ALC) mode ensures that the operating point of the HPA is independent of the uplink signal. This is best suited for a signal or a number of signals from a single location like a TV feeder link with several TV channels from the same teleport. It ensures that the

best HPA operating point is independent of the signal level at the input of the payload. This does not, however, ensure the best signal quality, since as the signal level at the input of the payload drops—because of, for example, rain fade—the uplink C/N drops but the downlink EIRP is maintained by the ALC and downlink C/N. The received C/N at Earth station will be affected by the degraded uplink C/N, although the net effect depends on the relative values of both C/Ns.

Of course, the payload is only a component of a communication system, and normally the ALC is employed in conjunction with uplink power control (ULPC) so that the feeder link or uplink station can increase the uplink power when there is fade to maintain a constant C/N at the satellite. However, RF power is a dimensioning parameter anywhere, including at the ground station, and there are limits. The limits of the ground station and the maximum fade determine when the flux density at the spacecraft can no longer be maintained and the payload ALC kicks in. In broadcasting satellites, the ULPC performance of the feeder link and the ALC performance of the payload are the main parameters defining the availability of the system.

The CAMP can also include a linearizing function that allows more power to be delivered from the TWTA for a given linearity. This type of CAMP, which is referred to as LCAMP, represents the general case today. For simplicity, since most CAMPs today are LCAMPs, the remainder of this book uses CAMP as the general term to include the linearizing function.

The TWTA is the main factor that determines the downlink quality or C/N. The TWTA incorporates the traveling wave tube (TWT) and its electronic power conditioner (EPC). The EPC provides all power lines to the TWTA and the CAMP but also includes the TC/TM (telecommand/telemetry) functions. TWTAs tend to be the most efficient HPAs with an efficiency of about 60%. The saturated RF of a TWTA is typically from 120-W RF to 200-W RF.

The CAMP, the TWTA, and the EPC go together. The linearizing function of the CAMP is customized to the specific TWT. Depending on the supplier, there could be three units (i.e., a CAMP, an EPC, and a TWT) or two units where the CAMP and the EPC are housed together. The latter is referred to as a microwave power module (MPM). Additionally, dual EPCs are also available so that two amplifying chains share a single EPC. Again, depending on the supplier, two chains may include two TWTs, two LCAMPs, and one EPC or two TWTs and an EPC that includes two CAMPs. The main benefit of dual EPCs is their mass saving and decrease in needed equipment, which has a bearing on the integration effort.

In addition to TWTs there are solid-state power amplifiers (SSPAs). Traditionally, they have been limited in the available RF power, which tends to diminish at higher frequency. Thus, at C-band 60-W RF may be available; this tapers to about a third at Ka-band. The efficiencies are also lower—about half that of TWTAs. SSPAs tend to be smaller and more reliable. Although their application in classical payloads has been limited, their utilization for more advanced architecture may be interesting.

The outputs the HPAs feed into the output multiplexer (OMUX). The function is reciprocal to that of the DEMUX, but the implementation is different. Considering that they are in the high-power section, low loss and high-power handling become important features. Waveguide technology is the mainstream, especially at the Ku-band and Ka-band. The number of poles tends to be four to six. Reducing the number of poles reduces the loss (and, of course, the filter performance). At C-band, dielectric technology may be used, but at higher frequencies either INVAR (for the required thermal stability) or aluminum (with compensation techniques) is used.

The combining system tends to be a waveguide manifold leading to a single output port that feeds the downlink antenna. The technology of the downlink antenna must be compatible with the downlink service area. Considering that this antenna carries several carriers, power handling is an important parameter.

Multipaction and passive intermodulation products (PIMPs) are also important considerations for both the OMUX and the antenna, because in addition to the total average RF power that the equipment sees, the number of carriers may lead to high peak power with its ensuing issues.

A payload can be described as having the following two systems:

- Antenna system;
- Repeater.

The repeater comprises the equipment between and including the LNA and the OMUX. Thus, the repeater input is the LNA input, and the repeater output is the OMUX output. The antenna system includes both the uplink antenna and the downlink antenna (illustrated in Figure 4.1).

A classical payload often uses the notion of a transponder. This is a chain of equipment that is required for the complete operation of a channel, as shown in Figure 4.2. Note that some equipment is required for a given transponder but is not unique to that transponder. Thus, an antenna, an LNA, and a downconverter are required for a transponder but are shared with other transponders. On the other hand, the filters are unique to the transponder, and a dedicated CAMP and a dedicated TWTA are also required for a given transponder.

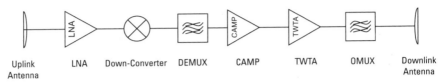

Figure 4.2 A transponder in a classical payload.

In general, classical payloads employ two polarizations on the uplink and two polarizations on the downlink. The uplink and downlink polarizations are orthogonal. This is illustrated in Figure 4.3, where the top half of the payload receives in linear polarization X and transmits in polarization Y and the lower half receives in Y and downlinks in X.

4.3   Current HTS Payloads

Current HTS payloads have evolved from the classical payloads and are based on similar equipment and technologies, although there have been developments for the specific HTS application. Of course, the HTS system is optimized

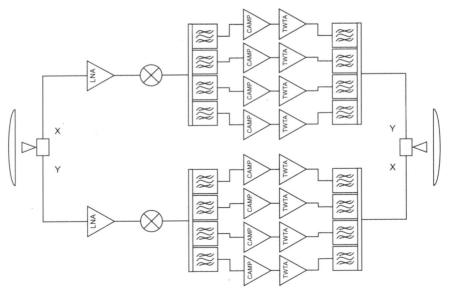

Figure 4.3 A functional block diagram of a dual-polarization classical payload.

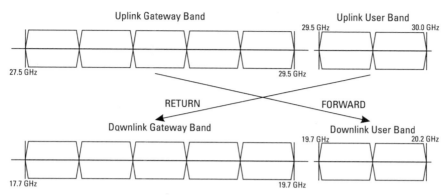

Figure 4.4 The basic frequency plan of a current HTS payload.

around the broadband access service, and the payload is an element of the system, which also includes the gateways and the terminals [1]. Figure 4.4 shows the basic HTS frequency plan [2]. In Figure 4.4, the following two links can be observed:

- The forward link communicates from the gateways to the users and therefore receives in the uplink gateway band and transmits in the downlink user band.
- The return link performs the reverse function, and its uplink uses the uplink user band and downlinks in the downlink gateway band.

In a current HTS system, the user band is very often in the Ka-band, and this is likely to continue. The gateways can also use the Ka-band but do not have to. Other bands like the Q/V-band and the W-band are available for the gateways.

Figure 4.5 shows a functional diagram of a current HTS payload, which functionally organized in the following two payloads:

- The forward payload;
- The return payload.

As can be seen in Figure 4.5(a), the forward payload has its uplinks from the gateways and the downlink to the user beams. In general, Ka-band HTS systems employ circular polarization for both the gateways and the user terminals. Obviously, this is reflected in the payload. In order to maximize gateway

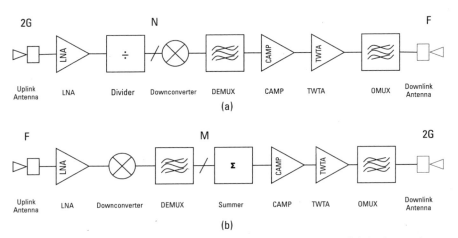

Figure 4.5 Functional block diagram of a current HTS payload with (a) the forward payload and (b) the return payload.

utilization, two polarizations are available per gateway, but one polarization is often used per user beam.

Assuming a transparent payload, the total bandwidth of the gateways must be equal to the total bandwidth of the users

$$\sum_{g=1}^{G} B_{g,\text{pol1}} + B_{g,\text{pol2}} = \sum_{g-1}^{G}\sum_{b=1}^{B} B_{bg} \qquad (4.1)$$

If we take KA-SAT as a basis with some simplifications, then $B_{g,\text{pol1}} = B_{g,\text{pol2}} = 2.5$ GHz, $B_{gb} = 250$ MHz, $G = 8$, and $B = 10$, with the total number of beams being $BG = 80$, so that the total forward system bandwidth is

$$\sum_{g=1}^{8}(2.5 \text{ GHz} + 2.5 \text{ GHz}) = \sum_{g=1}^{8}\sum_{b=1}^{10} 250 \text{ MHz} = 20 \text{ GHz} \qquad (4.2)$$

Looking at Figure 4.5, the number of uplink signals or frequency multiplexes is $2G$ where G is the number of gateways in the system operating on two polarizations. The gateway antenna feed could be shared with the user feeds if the gateways are embedded in the user service area, or they could be separate. In fact, the gateways could have a separate antenna system.

The uplink multiplexes are then amplified by the LNAs in the gateway uplink band, which, in Figure 4.5, would require $2G$ units, and downconverted to the user downlink band. The frequency conversion process is more complicated since the uplink gateway band is larger than the downlink user band. This follows the logic that it is desirable to minimize the number of gateways but also implies that one converter is inadequate. Referring to Figure 4.5(a) and using the simplified KA-SAT example, each gateway is associated with 10 beams. Assuming that both polarizations are used, this translates to two frequency multiplexes of five channels, one per beam. In this example, N would be equal to five so that five converters would be required per polarization. Thus the $2G$ uplink signal paths fans out to F, which is equal to $2GN$ paths after the conversion. In the KA-SAT example, $G = 8$, $N = 5$, and thus $F = 80$, which is the total number of beams. This is followed by appropriate filtering to suppress the adjacent channels and optimize the signal quality of the wanted channel.

The next step is high-power amplification to amplify the signals from the filters by the CAMP and to drive the TWTA. To ensure the best possible downlink signal quality, these amplifiers are typically operated in ALC mode. Thus, as rain fades occur on the uplink, initially the ULPC of the gateway kicks in trying to maintain the same power at the satellite uplink antenna. This also ensures a constant C/N at the satellite during ULPC operation.

However, RF power at the gateway is limited. Consequently, as the fade gets deeper, the uplink power hits the end stop, and no more power will be available from the gateway. At this stage, the signal level at the satellite starts dropping, as does the uplink C/N. This where ALC becomes active by introducing more gain to maintain the drive level of the TWTA constant. This is not ideal since the C/N of the input signal to the TWTA is degraded. However, if ALC were not to be used, there would a double hit: The downlink C/N would be at best equal to the degraded uplink C/N, but the TWTA would also deliver less power because of the reduced input drive, thereby reducing both the EIRP and the downlink C/N. The ALC thus maintains a constant EIRP with a degraded C/N that reflects the input degradation.

If the TWTA is carrying a single data stream that occupies the full beam channel, the amplifier can be operated at or close to saturation; this is ideal since it would attain the best spectral and power efficiency. However, the throughput of today's streams is a few hundreds of megabits per second, and the beam channel attains a gigahertz or more, so a number of streams are required to populate the channels. This leads to multicarrier operation, requiring a back-off in the TWTA. Typically, the back-off is between 2 and 3 dB, depending on the number of carriers. Since the throughput of streams is a function of technology, as technology advances and higher throughputs are achieved, the number of carriers decreases, and lower back-off can be applied. Of course, back-off is an operational condition, and as improved gateway equipment becomes available the satellite can adjust.

Finally, after high power amplification, the signals are filtered and fed to the corresponding ports of the beams in the antenna downlink system.

The return payload is functionally the reciprocal of the forward payload. For a given beam, the uplink and the downlink are orthogonal. As can be visualized in Figure 4.5(b), the signals from the respective beams are received by the antenna and appear at the corresponding ports of the uplink antenna system. The uplink user beams typically have the same geographic definition as the downlink user beams, and therefore there are F beams as in the forward link. The cell definition for the uplink is often identical to that of downlink.

In Figure 4.5(b), these signals are then amplified by F LNAs and fed to a filter multiplexer network. The function of this network is similar to that of an OMUX in a classical payload, but the technology is different since there are no high-power requirements. In Figure 4.5(b), this is represented as filters followed by a summer so that M channels are multiplexed together to form a single signal structure.

In the KA-SAT example, five channels are combined to form the multiplex for a given polarization for a given gateway so that $M = 5$, and there are F/M outputs, which is 16 and equal to $2G$, the number of gateways employing two polarizations. The combined signal is then fed to the CAMP. Since the combined signal is made up of several carriers from each beam and of the composite signals from a number of beams, the FGM mode of the CAMP is selected; otherwise the downlink EIRP per carrier would be a function of the total activity of the carriers.

The CAMP then drives the TWTAs. These amplifiers operate in a multicarrier mode and therefore operate in back-off, typically 4 dB or more. The fact that a single TWTA is used for a large number of carriers from a number of beams is a significant economic factor of HTS systems since the return capacity is considerably less expensive in terms of the numbers of TWTAs, the direct current (DC) power requirement, and the spacecraft dry mass.

Comparing Figures 4.3 and 4.5 shows that although most of the functional blocks are the same, the topologies of the payloads bear differences, described as follows.

- The air interfaces of the user links of an HTS are normally better than those of an equivalent classical payload. This is a key design feature.
- The air interfaces of the gateway links of an HTS are normally worse than those of an equivalent classical payload. This reduces the demand for resources of the payload while taking advantage of the size of the gateways
- The air interfaces of the classical payload are normally more balanced between the uplink and downlink, because often the classical payload supports terminal-to-terminal communications. This is not necessarily

the case in a broadcast satellite where the uplink performance can be relaxed.

- Single-hop meshed communications are supported at the physical level in a classical payload. Meshed communications are supported at the network level and require a double hop.
- HTS payloads employ circular beams while classical payloads utilize contoured beams.
- The air interface of a classical payload is to some extent compromised due to the large antenna coverage and the ensuing lower antenna gain.
- The notion of a payload per polarization disappears in the HTS payload. In fact, orthogonal polarization for the uplink and downlink signals is not required for the forward payload or the return payload.
- The topology of the HTS payload is implicitly unidirectional. Bidirectional communications are achieved by two payloads: the forward payload and the return payload.
- The topology of the HTS payload is implicitly a star configuration: a gateway reaching out to a number of beams on the forward link and a number of beams feeding into a gateway for the return link.
- The notion of a transponder is less evident as a commercial unit for an HTS system, and capacity becomes the commercial unit.

4.4 System Aspects/Trade-Offs

The spacecraft antenna system of an HTS is a dimensioning factor. There are two basic reflector technologies available: solid reflectors and mesh reflectors. Practically, all current HTSs use solid reflectors. This is because the technology is very mature, but it limits the size to about 2.5m in diameter. This physical aperture diameter translates to a full beamwidth of 0.4°.

4.4.1 Color Schemes: Frequency, Antennas

The most common usage of the color scheme is in terms of the frequency plan. The purpose is to define a set of spectral resources that are mutually orthogonal (i.e., do not overlap in frequency and/or are on different polarizations). These spectral resources can then be reused, leading to frequency reuse and the resulting spectral efficiency.

Using KA-SAT as the example, the user band is 500 MHz on both the uplink and the downlink. Each spectrum is divided in two on each polarization so that there are four 250-MHz channels. Each channel is assigned a color so that all beams using the same color would frequency-reuse the assigned channel. This describes a homogenous system, since the assumption is that all beams have the same demand with all beams having 250 MHz assigned to them. Therefore, by definition, this is an even-color scheme. This is illustrated in Figure 4.6(a).

The demand over the service area is rarely homogeneous; therefore the demand per beam can be quantized in two sets of bandwidths that fit in the spectrum on the two polarizations, and this can be used to define the color scheme. The number in the sets should be small to maximize the frequency reuse, but the number of bandwidths in each set (or polarization) need not

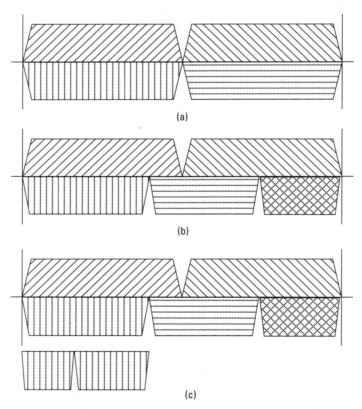

Figure 4.6   Examples of types of user frequency plans that can be employed in a HTS system (uplink or downlink): (a) a frequency plan employing a four-color scheme for a homogeneous system with all colors having the same bandwidth, (b) a frequency plan employing a five-color scheme with two colors having the same bandwidth assigned to one polarization and three colors with different bandwidths assigned to the orthogonal polarization, and (c) which is identical to (b) except for that one color has the option of being split into two subcolors.

be the same. Therefore, the resultant frequency scheme does not need not be as even as shown in Figure 4.6(b).

Subcolors are also a possibility. This is where a color can either be used as is or split into components. In Figure 4.6(c) a channel can either be used as such or as two channels. This provides flexibility in the apportionment of resources without increasing the resources. For example, for a given system with a given number of TWTAs, the system bandwidth is determined. Thus, with the same number of TWTAs, the service area can be extended.

The antenna system needs to generate the lattice of beams that constitutes the service area while maintaining antenna efficiency and minimizing the intrasystem interference due to the interbeam interference. The antenna system is typically a reflector antenna. Thus, if a single beam is generated employing a dedicated feed, the dimensions of the feed are optimized so that given the focal distance/diameter ratio (f/d), most of the radiation pattern of the feed is captured by the reflector. This reduces the spillover loss of the antenna and increases antenna efficiency. It also reduces the sidelobe level, which is desirable to control the intrasystem interference. This generally implies a large feed, which is not compatible with the beam spacing. In a single-feed-per-beam (SFPB) antenna system, each beam maps to a single feed. Typically, each feed is used on both polarizations: one for the uplink and the other for the downlink. As the feed becomes large, the interfeed dimensions increase, and consequently the beam spacing increases.

Employing a reflector color scheme is a solution. Figure 4.7 illustrates an implementation for the service area discussed in Chapter 2, employing a four-color scheme for the antenna system. The lattice of beams is organized into four colors (e.g., red, blue, green, and yellow). The objective is to maximize the spacing of the cells of a given color. Notice that Figure 4.7 shows hexagons, not circles. These are the cells. A user must be associated at any given time to a cell even though it may be covered by two beams. A cell can be defined as the part of the beam coverage that is uniquely used as a part of the service area.

The composite lattice, which is predominantly contiguous depending on the service area, is decomposed into four lattices that are sparser, designated as the red, blue, green, and yellow lattice. The sparsity eases the feed design and enhances the performance of the antenna. The implementation of such a system could be a four-reflector antenna system where each feed system is a mapping of the corresponding cell lattice as shown in Figure 4.7. This means that the lattice of cells corresponding to the service area is generated by four-antenna systems.

It should be noted that when both the antenna and the frequency plan employ a four-color scheme, the antenna color scheme does not have to be

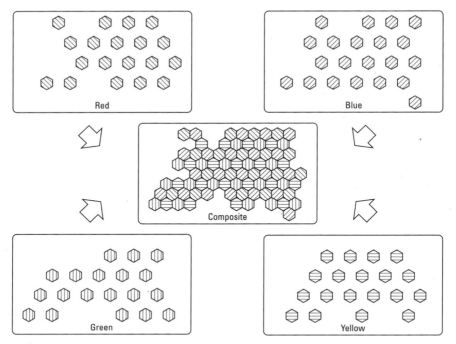

Figure 4.7   A four-color scheme employed by the antenna system, showing the component color sets of cells and the composite lattice of cells.

aligned with the spectral color scheme, although it is often convenient to have a four-color scheme be the basis for both.

A four-color antenna system consists of four large reflectors, which are demanding in terms of real estate. To reduce the accommodation requirements, a three-color reflector scheme can be used. Three-antenna systems, like the one shown in Figure 4.8, use three reflectors (red, blue, and green in Figure 4.8). Similarly, the contiguous lattice corresponding to the service area is split into three sparse lattices (instead of four). The ensuing interbeam spacing is less than that of the four-color scheme, but it may be acceptable. There are two effects, as the feeds in a three color-scheme are smaller than those in a four-color scheme, the antenna gain decreases as the spillover loss increases and the higher side-lobes results due to an increase is the edge taper. Additionally, in three-antenna systems, since the cells of a given color are physically closer, neighboring cells are higher up the sidelobe slope.

The antenna gain could be counteracted through an oversized reflector to diminish the direct impact on capacity, but this is not entirely the case for the C/I. In the forward link budget, the C/I is a function of the following:

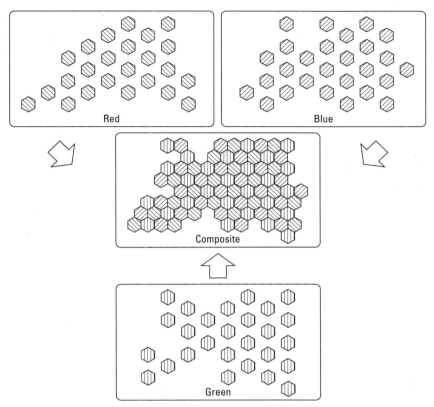

Figure 4.8 A three-color scheme employed by an antenna system, showing the component color sets of cells and the composite lattice of cells.

- The cell isolation which result from neighboring cells operating at the same frequency and with the same polarizations;
- The cross-polar discrimination which result from neighboring cells operating at the same frequency and with the orthogonal polarizations.
- The carrier-to-intermodulation ratio when in multicarrier operation;
- The adjacent channel interference from other cells;
- The downlink of the return link, if the return downlink band is adjacent to the forward downlink.

The ultimate objective of an HTS system is to maximize the capacity. Finding an adequate antenna system requires a complex and holistic approach that weighs the antennas' overall impact on the system capacity against the system cost.

4.4.2 Beam Size Versus System Complexity

In Chapter 2, the relationship between the number of beams required to populate a given service area was given as

$$N = \frac{\alpha_{SA}\beta_{SA}}{k_{cell}\theta_b^2} \qquad (4.3)$$

Thus, if the bandwidth assigned to the ith cell is b_i

$$B_{sys} = \sum_{i=1}^{N} b_i \qquad (4.4)$$

To simplify the analysis the average bandwidth per cell could be used with the ensuing simplification

$$B_{sys} = N\overline{b} \qquad (4.5)$$

In a homogeneous system, b_i and $\overline{b}$ are equal.

For the forward payload, if the service requirements specify an average EIRP spectral density of $\overline{eirp}$ then,

$$\overline{eirp} = \frac{P_{RF,sys}\,\overline{G_{ant}}}{B_{sys}} \qquad (4.6)$$

where $P_{RF,sys}$ is the total RF power delivered to the antenna system, and $\overline{G_{ant}}$ is the average antenna gain over the service area.

Equation (4.6) can also be written as

$$EIRP_{sys} = P_{RF,sys}\,\overline{G_{ant}} \qquad (4.7)$$

The system EIRP, $EIRP_{sys}$, is rather conceptual and a system parameter that could be useful for payload-dimensioning but could also be used with caution in link budgets[1]

The average EIRP in a given cell, can be written as

---

[1] The system EIRP is the aggregate of all the EIRPs on all coverages. It is a metric of the size of the HPA system and the associated antennas. The size of the HPA system has a significant impact on the spacecraft power subsystem, and thus the size of the HPA system, the antenna farm, and the spacecraft power subsystem together dimension the spacecraft to a large extent. However, it is not available at any point in the coverage and, therefore, cannot be used directly in a link budget.

$$\text{EIRP}_{\text{cell}} = \frac{P_{\text{RF,sys}} \overline{G_{\text{ant}}}}{N} \quad (4.8)$$

Now the beamwidth of the antenna can be expressed as

$$\theta_b = \frac{k_{\text{ant}} \lambda}{D_{\text{ant}}} \quad (4.9)$$

where:

k_{ant} is a constant;

λ is the wavelength;

D_{ant} is the diameter of the reflector aperture.

By convention, the beamwidth is expressed in terms of the 3-dB beamwidth, and in this case k_{ant} is typically 70. However, this is not the only possibility, and if the beamwidth is defined with respect to the 4-dB beamwidth, k_{ant} would be about 81.

Thus, the peak antenna gain can be expressed with the beamwidth as the variable as

$$G_{\text{ant}} = \frac{\eta_{\text{ant}} \pi^2 k_{\text{ant}}^2}{\theta_b^2} \quad (4.10)$$

where η_{ant} is the antenna efficiency.

At the system level, the antenna efficiency can be interpreted as the directivity-to-gain factor.

The average antenna gain can be derived from the peak antenna gain and given as

$$\overline{G_{\text{ant}}} = \frac{\eta_{\text{ant}} \pi^2 k_{\text{ant}}'^2}{\theta_b^2} \quad (4.11)$$

where k_{ant}' is the modified constant to give the average antenna gain.

Replacing the average antenna gain given in (4.11) in (4.8), we get

$$\text{EIRP}_{\text{cell}} = \frac{P_{\text{RF,sys}}}{N} \frac{\eta \pi^2 k_{\text{ant}}^2}{\theta_b^2} \quad (4.12)$$

Now relating the beam size to the service area as given in (4.3), we obtain

$$\text{EIRP}_{\text{cell}} = \frac{\eta \pi^2 k_{\text{ant}}^2 k_{\text{cell}} P_{\text{RF,sys}}}{\alpha_{SA} \beta_{sa}} \qquad (4.13)$$

This is an interesting result since given the size of the service area $(\alpha_{SA}\beta_{SA})$, the antenna technology (k_{ant}), and the cell form (k_{cell}), the EIRP in a given cell is independent of the number of cells (N). This is not entirely surprising since it is the definition of EIRP, but in an HTS context, it means that the capacity density can be increased without necessarily increasing the required RF power, assuming that the power can be apportioned between the cells. Small cells offer a bigger potential for the management of RF power to meet the demand over the service area in a dynamic fashion. As the antenna size is increased, the required RF power per beam is decreased and applying the previous observation of constant system RF power, more users can be served. Of course, this comes at the cost of a larger antenna and system complexity since the RF power must be managed over a larger number of ports. This observation is particularly useful when looking at systems that are not homogeneous in the geographic spread of capacity requirements.

The bandwidth that is available in a cell is limited by the available spectrum. In the Ka-band in Region 1 as defined by the ITU this could be 17.3–20.2 GHz (i.e., 2.5 GHz). It is possible to assign the full spectrum to a cell, but this causes constraints on the adjacent cells. Defining the available spectrum as B_{av}, we can say that $0 \le b_i \le B_{av}$. We can assume that the system RF is derived from M HPAs and that each HPA has an operational bandwidth of B_{HPA} so that

$$\frac{N}{M} = \frac{B_{\text{HPA}}}{b} \qquad (4.14)$$

Today the HPA is typically a TWTA, but it could also be a SSPA. If we assume a four-color scheme for the frequency plan, the maximum capacity that the system can deliver is when $\bar{b} = 1/2 B_{av}$, and the required condition to minimize the number of TWTAs is $B_{\text{TWTA}} = B_{av}$. Under these conditions, each HPA can be used to drive two beams. Of course, the maximum capacity requirement over the service area may not necessarily be a requirement as although it may maximize the system capacity over the service area, it may be putting capacity where it is not required. This may not reflect the commercial requirement. In this case, an HPA can sometimes be used to drive more than two beams as illustrated in Figure 4.6(c).

This discussion demonstrates that there are three desirable requirements for the HPA, described as follows:

- A requirement for the operational bandwidth, B_{HPA}, to be equal to the available bandwidth, B_{av};
- An HPA operational RF that ensures the required EIRP density over the operational bandwidth;
- Good efficiency at the HPA operating point over the operational bandwidth.

This leads to an optimization in the topology and a reduction in the equipment count since the topology shown in Figure 4.5 can be modified. In the KA-SAT example, a TWTA operating over the required 500 MHz can be used per beam pair, with each using 250 MHz. Thus, with 80 equivalent beams, 40 active TWTAs are utilized. Recall that this is where efficiency is important in terms of simplifying the payload topology and consequently the mass (and equipment count) and cost. Further, the use of a TWTA pair with a common dual EPC also adds to the mass efficiency. Again, in the KA-SAT example, 20 active TWTA pairs with dual EPC can be used.

This topology requires new equipment. The OMUX is no longer a single filter per TWTA. It is a two-filter unit, sometimes referred to as a reverse OMUX. In fact, although the name OMUX is used due to its location in the payload topology and its function is that of a DEMUX as it separates the composite spectrum amplified by the TWTA into the spectra intended for the specific beams. It should be noted that the total number of filters remains unchanged although the number of units is halved.

This topology has its own challenges. The TWTA amplifies a number of signals intended to different beams. This could lead to a multipath issue. If the terminal receives the signals amplified by the same amplifier and therefore coherent versions of the same signals from different beams, the signals will add vectorially at the terminal with the resulting ripple in the signal. The effect is worse when the signal amplitudes are close to each other. The worst case would exist if the terminal is at the edge of a cell and the adjacent cell is amplified by the same amplifier when using frequencies that are close to those of the adjacent cell and copolar.

This can be mitigated in the following three ways:

- Frequency;
- Spatial separation;
- Polarization.

If terminals at the intersecting edge use frequencies that are far away from the frequencies used by the adjacent cell, the filter performance of the adjacent cell is better and consequently will have better suppressed the unwanted multipath signal, reducing the ripple. This is an operational workaround solution and a constraint. In general, operational constraints are not desirable and should be avoided if possible.

If beams that are amplified by the same amplifier are remote within the service area, the antenna gain associated with the unwanted multipath signal decreases with spatial separation, and therefore the multipath effect is reduced [3]. It also complicates the waveguide routing between the repeater and the antenna. The ease of finding a solution for beams associated with a given TWTA increases with a large system.

Similarly, polarization can also be used to suppress the multipath effect. If the TWTA is used to amplify signals on opposite polarizations, the terminal will receive the multipath signal of the orthogonal polarization, and therefore the ripple effect will be reduced accordingly [3].

The three methods could be used together to increase the multipath suppression, although typically with a good payload design, operational constraints can be eliminated or at least contained to limited areas.

Further optimization is still possible to reduce equipment count. Using the KA-SAT example, each gateway is associated with 10 beams, five on each polarization. Similar to the TWTA case, channel pairs can use a converter to translate an adjacent channel pair together, thus reducing the number of operational converters required. In the KA-SAT example, three converters are required per polarization instead of five.

Thus, the conversion is typically performed in groups of channels that fully occupy the available downlink user band. It should be noted that the organization of the channels on the uplink has little technical impact and is open to simplify the conversion process. This makes it possible to use a converter per two (or more) channels so that the number of converters for the forward payload is also significantly reduced.

Figure 4.9 illustrates optimization in the payload topology. In Figure 4.9(a), the number of feeds for the uplink and the downlink remain unchanged (i.e., $2G$ and F, respectively). These are derived from a higher-level requirement. However, N is modified to N', which in the KA-SAT case would be 5 and 3, respectively, corresponding to a savings of 40% of the converters. It could be 50% if N were to be paired. The savings in the TWTA is 50%, since unpaired outputs with nonoverlapping spectra from the DEMUXs can be paired to the CAMP input.

Optimization on the return payload is also possible, and the following two cases can be identified:

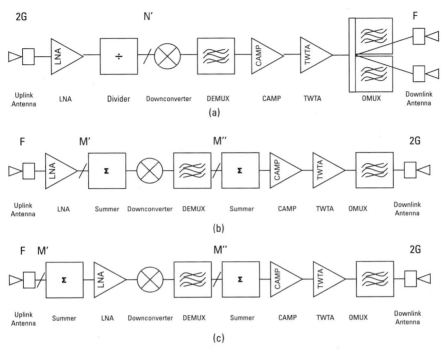

Figure 4.9   Optimization in the payload topology for the forward and return sections. (a) shows an optimization for the forward payload where N'<N reducing the number of embarked unis for the downconverters, CAMPs, TWTAs, and OMUX units. (b) and (c) show optimizations for the return payload. With M'<M and M''<M the number of embarked downconvertors, CAMPs, TWTAs, and OMUX filters are reduced.

- A case where the return link budget is interference-limited;
- A case where the return link budget is noise-limited.

Obviously in both cases, the number of feeds on the uplink and the downlink remain unchanged since this is a higher-level requirement. Referring to Figures 4.5 and 4.9 these are *F* and 2*G*, respectively.

In a noise-limited case, it is important to conserve the *G/T* of the return payload, and therefore an LNA following the antenna is essential in order not to degrade the system performance. The number of LNAs is, thus, unchanged and maintained at *F* as can be seen in Figure 4.9(b). The signals are combined post-LNA. Of course, the signal combining as described needs to ensure that the spectra being combined do not overlap.

As the antenna gain increases, the G/T increases to a level that cannot be operationally exploited since the link budget becomes swamped by other factors like interference. When the return link is interference-limited, degradation in the G/T becomes acceptable, and the combiners at the antenna port become an attractive solution since they reduce the number of LNAs. This is illustrated in Figure 4.9(c). In the KA-SAT example, this would represent a savings of 50%, from 80 to 40, and M' is 2.

In both cases, the combined spectrum is then converted to the downlink frequency with the same savings due to M'. The savings is not just in number of convertors but also in converter types. Referring to KA-SAT, instead of five types, we would have three types. The ensuing spectra is then summed again to constitute the signal structure per polarization to the gateways (i.e., $2G$ composite signal structures intended for both polarizations of the gateways).

References

[1]   Fenech, H., et al., "VHTS Systems: Requirements and Evolution," *11th European Conference on Antennas and Propagation (EuCAP)*, Davos, Switzerland, March 20–24, 2017.

[2]   Fenech, H., "KA-SAT and Future HTS Systems," *The Plenary Session of the 14th IEEE International Vacuum Electronics Conference (IVEC)*, May 21–23, 2013, Paris, France.

[3]   Fenech, H., and E. Lance, *Payload for Multi-Beam Satellite*, French patent: FR2950496 (B1), granted October 21, 2011; U.S. patent: US9118384 (B2), granted August 25, 2015.

5

HTS System Analysis: The Forward Link

5.1 Overview

This chapter analyzes the salient parameters that define the HTS system and its capacity for the forward link [1]. The analysis is limited to a theoretical system that greatly simplifies the work but remains still relevant as it gives good insight into real systems. To the extent possible, the chapter references real systems to explain deviations.

Section 5.2 examines the spacecraft antenna, using a theoretical model of an antenna to highlight the key elements that impact system performance. Section 5.3 develops the forward EIRP model over the area of a cell. Section 5.4 studies the forward downlink interference model, which employs simulation techniques with a theoretical antenna system to study the interference scenario over a cell. Other spacecraft impairments that could have an impact on the link budget and consequently the capacity are covered in Section 5.5. Recall that a key enabler in HTS systems is adaptive coding and modulation; accordingly, Section 5.6 adopts and reviews the DVB-S2x as a reference. Finally, Section 5.7 discusses the link budget to estimate the capacity of a cell and the dependence on system bandwidth.

5.2   The Antenna

The antenna is a key element in an HTS system, not only because it defines the service area but also because it has a strong bearing on the capacity. The characterization of antenna can be complex, but this section limits the analysis to the features that are important at the system level for an HTS system.

For HTS systems to obtain good performance, the antennas should have narrow beams with fast roll-offs and low sidelobes for achieving good C/I. These beams are sometimes referred to as pencil beams. As described in [2, 3], when considering circular apertures, the beamwidths and sidelobes are a trade-off.

For a circular aperture, the field illumination of the reflector by the feed can be represented by a parabolic profile as

$$E_a(\rho') = C_{ant} + \left(1 - C_{ant}\right)\left[1 - \left(\frac{\rho'}{a}\right)^2\right]^n \tag{5.1}$$

where:

C_{ant} is the pedestal level $0 \leq C_{ant} < 1$;

ρ' is the radial distance from the center of the aperture;

a is the radius of the aperture;

n is the order of the illumination profile;

$n = 0$ for a uniform illumination, and the function is independent of the edge taper;

$n = 1$ for a parabolic illumination;

$n = 2$ for a squared parabolic illumination.

The family of illumination patterns are plotted in Figure 5.1 for a uniform ($n = 0$), parabolic ($n = 1$) and squared parabolic ($n = 2$) distribution and for an edge taper of 0.316 (10 dB) and 0.1 (20 dB). Although the edge taper is a negative number in decibels, the negative sign is often omitted.

The normalized pattern resulting in the far-field of such an antenna configuration, is given by [2, 3]

$$F\left(u, n, C_{ant}\right) = \frac{C_{ant} f(u, 0) + \dfrac{\left(1 - C_{ant}\right)}{\left(n + 1\right)} f(u, n)}{C_{ant} + \dfrac{\left(1 - C_{ant}\right)}{\left(n + 1\right)}} \tag{5.2}$$

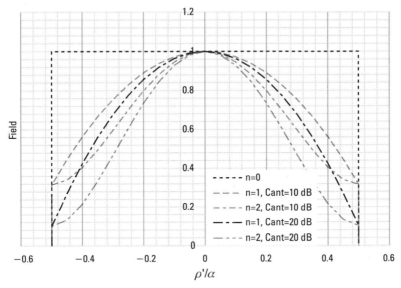

Figure 5.1 Reflector illumination employing (5.1) for a uniform ($n = 0$), parabolic ($n = 1$) and squared parabolic ($n = 2$) distribution and for an edge taper of 0.316 (10 dB) and 0.1 (20 dB).

where

$$f(u,n) = 2^{n+1}(n+1)! \frac{J_{n+1}(u)}{u^{n+1}}$$

and

$$u = \frac{2\pi a}{\lambda}\sin\theta$$

where $J_n(x)$ is a Bessel function of the first kind and of the nth order.

The function $F(u,n,C_{\text{ant}})$ represents the far field so that the directivity can be expressed as

$$D(u,n,C_{\text{ant}}) = \left|F(u,n,C_{\text{ant}})\right|^2 \tag{5.3}$$

Figure 5.2 plots (5.3) for a center-feed antenna with a diameter of 2.6m at 19.7 GHz for $n = 0$, 1 and 2 for an edge taper of 10 dB and 20 dB. Two regimes can be recognized in the pattern: the main lobe and the sidelobe

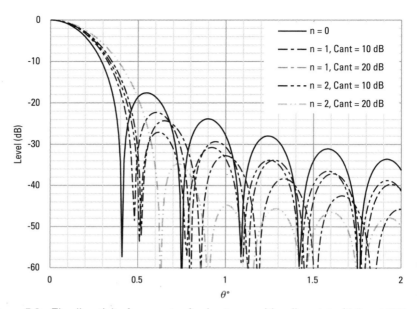

Figure 5.2   The directivity for a center-feed antenna with a diameter of 2.6m at 19.7 GHz for a parabolic on pedestal illumination for $n = 0$, 1 and 2 and for an edge taper of 10 dB and 20 dB.

structure. The optimal antenna would maximize the directivity over a small circular area while suppressing the sidelobes to an adequate level.

It will be noticed that when $n = 0$, the main lobe is the most compact and the half-power beamwidth (HPBW), θ_{3dB} is the smallest, and thus the peak directivity is the highest—although this cannot be seen in Figure 5.2 because of the normalization of the plots. The aperture is uniformly illuminated and therefore fully utilized to the maximum. However, there is a severe discontinuity at the aperture rim, which produces high sidelobe levels. This is illustrated in Figure 5.2 and Table 5.1.

As the order of the illumination increases and/or the edge taper decreases, the main lobe becomes broader and the peak directivity decreases. This is because the aperture although fully illuminated, is not utilized to the same extent on a radial cut. This effect can be termed as the illumination efficiency. Additionally, although the illumination profile discussed above assumes that the pattern is as such, they do not exist in real life. The pedestal indicates that the illumination extends beyond the reflector. The reflector is not capturing the full illumination profile from the feed. Some of the energy is falling beyond the aperture and this is the spillover loss. As the edge taper increases (in positive

Table 5.1
The 3-dB Beamwidth, the First Sidelobe Level, and the Relative Level of the Main Lobe for a Center-Fed Circular Aperture with a Parabolic on Pedestal Illumination for $n = 0$, 1, and 2 and for an Edge Taper of 10 dB and 20 dB [3]

n	C_{ant} (dB)	θ_{3dB}	Sidelobe Level (dB)	Relative Directivity (dB)
0	—	$58.4\lambda/2a$	−17.6	0
1	10	$65.3\lambda/2a$	−22.3	−0.38
2	10	$67.0\lambda/2a$	−27.0	−0.57
1	20	$69.3\lambda/2a$	−24.3	−0.88
2	20	$75.6\lambda/2a$	−34.7	−1.61

dB), the discontinuity at the aperture rim is softened and this yields lower side-lobes as can be seen in Figure 5.2 and Table 5.1.

For a given aperture, the edge taper is a function of the feed pattern mainly related to its size. Although our discussion assumes a center-fed antenna configuration, this is rarely the case for a spacecraft antenna where offset fed antennas are common. This together with the fact that the antenna is not freestanding but mounted on a sidewall of the spacecraft complicates the situation and the analysis. While this discussion does not consider these factors, it is nevertheless, a basis to provide insight into the key system parameters or to provide a framework for a system analysis in which the specific antenna details are available.

The main lobe is of primary interest for the EIRP over the cell. The 3-dB beamwidth can be approximated as given in [3].

$$\theta_{3dB} = \frac{k_{ant}\lambda}{2a} \tag{5.4}$$

where $k_{ant} = \alpha_{ant}C_{ant,dB} + \beta_{ant}$

Table 5.2 provides the values for the parametric constants for (5.4).

For the main lobe of an axisymmetric antenna, a suitable simplification for the antenna directivity with respect to the 3-dB beamwidth in terms of the off-axis angle, θ, is given as

$$D_\theta = D_{pk} - 12\left(\frac{\theta}{\theta_{3dB}}\right)^2 \tag{5.5}$$

where θ_{3dB} is the HPBW in degrees, and D_{pk} is the peak directivity in decibels.

Table 5.2
Parameters for an Approximation of the 3-dB beamwidth for a Center-Fed Circular
Aperture with a Parabolic on Pedestal Illumination for $n = 0$, 1, and 2 [3]

	$n = 0$	$n = 1$	$n = 2$
β_{ant}	58.4	60.7	58.4
α_{ant}	0	−0.458	−0.859

Equations (5.2) and (5.5) were compared over the range of the peak directivity to −6 dB. Errors of up to 0.48 dB were found over the range of parameters and the directivity range. The errors increase closer to the −6-dB level. However, a modified value of θ_{3dB} can be determined to minimize the error to about 0.18 dB. This confirms that the simple form of the equation as defined in (5.5) is a sound basis for the analysis and provides a representative function that is more easily manageable mathematically. For the sake of simplicity, θ_{3dB} will be used for the rest of the discussion even though this may be modified.

As the 3-dB contour is not necessarily used as the eoc contour, the equation is generalized to

$$D_\theta = D_{pk} - 4\Delta D_{eoc} \left(\frac{\theta}{\theta_{eoc}} \right)^2 \tag{5.6}$$

where ΔD_{eoc} is the normalized level (positive in decibels) for the defined eoc contour, and θ_{eoc} is the full beamwidth at the eoc in degrees.

Note that θ is a one-sided, half-angle, or off-axis angle while θ_{3dB} and θ_{eoc} are full or double-sided angles.

Thus, although the illumination patterns considered are mostly theocratical, an insight is provided into how it characterized the beam in terms of beamwidth, gain and side-lobe performance.

5.3 The User EIRP Model

The capacity of an HTS system is closely related to the EIRP, and the spacecraft is, to a large extent, dimensioned by the power subsystem with its batteries and the solar arrays. The RF power system takes a large portion of this power. Here we derive the EIRP relationship and work out the EIRP profile across a theoretical hexagonal cell.

The EIRP in a given direction is given as

$$\text{EIRP}_{\theta,\varphi} = P_{\text{TWT}} - L_{\text{op}} - L_{\text{ant}} + D_{\theta,\varphi} \tag{5.7}$$

where:

P_{TWT} is the useful TWT RF power in dBW into the port of the antenna generating the beam referred to the TWTA output;

L_{op} is the repeater output losses in decibels of the associated payload path;

L_{ant} is the antenna losses in decibels of the given beam;

$D_{\theta,\varphi}$ is the antenna directivity of the given beam in the given direction.

This can be elaborated by putting (5.2) in (5.3). However, for the EIRP of an HTS, it is convenient to consider a cell as a hexagonal surface that is completely enclosed by a circle corresponding to the eoc contour of the specific beam as shown in Figure 5.3. Thus, assuming concentric alignment of

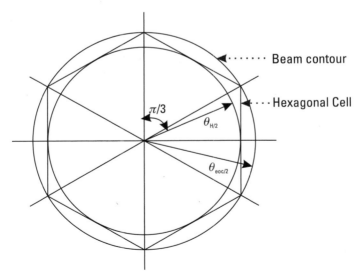

Figure 5.3 Circular beam with a beamwidth of θ_{eoc} covering a hexagonal cell.

the beam and the cell and the simplified EIRP expression from (5.6), we can express the EIRP as

$$\text{EIRP}_\theta = \text{EIRP}_{Pk} - 4\Delta D_{eoc}\left(\frac{\theta}{\theta_{eoc}}\right)^2 \tag{5.8}$$

where:

EIRP_θ is the EIRP in dBW at an off-axis angle θ of a given beam;

EIRP_{Pk} is the peak EIRP in dBW at the boresight of that beam and the cell since both are assumed to concentric;

θ_{eoc} is the full eoc beamwidth in degrees.

Note that since an axisymmetric beam is assumed, the EIRP can be expressed with a single variable rather than two as given in the general expression of (5.7).

Expressing the EIRP as a function of the area to be covered, the cumulative EIRP function can be obtained. The following two regions can be identified:

- Region 1: The coverage within the circle with diameter θ_H that is circumscribed by the hexagonal cell.
- Region 2: The remaining coverage outside region 1 but within the hexagonal cell.

The area within region 1 can be expressed in terms of the radius θ as

$$A_{\theta,R1} = \pi\theta^2 \tag{5.9}$$

θ_H can be expressed in terms of θ_{eoc} as

$$\theta_H = \frac{\theta_{eoc}\sqrt{3}}{2} \tag{5.10}$$

Thus, the total area in region 1 is given as

$$A_{\theta_H} = \frac{\pi\theta_H^2}{4} = \frac{3\pi\theta_{eoc}^2}{16} \tag{5.11}$$

while the area covered by the cell is given as

$$A_{\text{cell}} = \frac{3\theta_{eoc}^2\sqrt{3}}{8} \tag{5.12}$$

We can normalize the area with radius θ in region 1 with respect to that of the cell to get

$$a_{\theta,R1} = \frac{8\pi}{3\sqrt{3}}\left(\frac{\theta}{\theta_{eoc}}\right)^2$$

$$\text{For } 0 \le \theta \le \frac{\theta_H}{2} \text{ or } 0 \le \left(\frac{\theta}{\theta_{eoc}}\right)^2 \le \frac{3}{16}$$

(5.13)

A normalized variable can be defined where $\theta' = \theta/\theta_{eoc}$

$$a_{\theta,R1} = \frac{8\pi}{3\sqrt{3}}\theta'^2$$

$$\text{For } 0 \le \theta' \le \frac{\sqrt{3}}{4}$$

(5.14)

The normalized area for region 1 can be obtained either by dividing (5.11) by (5.12) or by putting $\theta = \theta_H/2$ in (5.14). Both yield the result as

$$a_{\theta_H,R1} = \frac{\pi}{2\sqrt{3}} = 90.7\%$$

(5.15)

Thus, using (5.8) and (5.11), the EIRP can be expressed in region 1 as a function of the normalized cell area as

$$\text{EIRP}_\theta = \text{EIRP}_{Pk} - \Delta D_{eoc}\frac{3\sqrt{3}}{2\pi}a_{\theta,R1}$$

$$\text{For } 0 \le a_{\theta,R1} \le a_H$$

(5.16)

The choice of ΔD_{eoc} is the result of a trade-off resulting from the antenna optimization within the system confines.

At the limit of region 1 (i.e., at $\theta = \theta_H/2$),

$$\text{EIRP}_H = \text{EIRP}_{Pk} - \frac{3}{4}\Delta D_{eoc}$$

(5.17)

Thus, in region 1 (i.e., $0 \le \theta \le \theta_H/2$), the EIRP decreases linearly from EIRP_{Pk} to EIRP_H a level that is $(3/4)\Delta D_{eoc}$ below the peak.

Looking at region 2, it should be clear that going from $\theta_H/2$ to $\theta_{eoc}/2$ the EIRP level drops another $\Delta D_{eoc}/4$. The geometry of region 2 is a bit more

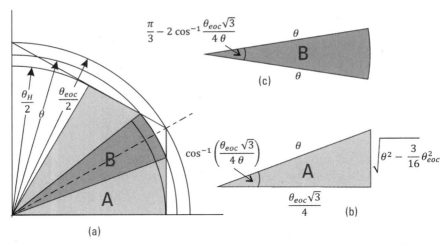

Figure 5.4   Details of the sectors defining one of the six identical sectors of region 2.

complicated. By visual inspection, it can be noted that the area consists of six identical sectors and that each sector can be further split into three sectors: area A, area B, and area C as shown in Figures 5.3 and 5.4(a). It should be evident that areas A and C are symmetric and identical in area.

Area A is a right-angle triangle with the dimensions given in Figure 5.4(b). Area B is a segment as illustrated in Figure 5.4 (c). Using the symmetry of the hexagonal structure, the area of region 2 can now be expressed as

$$A_{\theta,R2} = 12(\text{Area A}) + 6(\text{Area B}) \tag{5.18}$$

Applying appropriate geometry as illustrated in Figure 5.4, the area of region 2 can now be expressed as

$$
\begin{aligned}
A_{\theta,R2} &= \frac{12}{2}\left(\sqrt{\theta^2 - \frac{3\theta_{eoc}^2}{16}}\,\frac{\sqrt{3}\theta_{eoc}}{4}\right) + \frac{6}{2}\left(\frac{\pi}{3} - 2\cos^{-1}\frac{\sqrt{3}\theta_{eoc}}{4\theta}\right)\theta^2 \\
&= \frac{3\sqrt{3}\theta_{eoc}}{2}\sqrt{\theta^2 - \frac{3\theta_{eoc}^2}{16}} + \pi\theta^2 - 6\theta^2\cos^{-1}\frac{\sqrt{3}\theta_{eoc}}{4\theta}
\end{aligned}
\tag{5.19}
$$

For $\theta_H \geq 2\theta \geq \theta_{eoc}$

Noting that the total area of the cell is $(3\sqrt{3}/8)\theta_{eoc}^2$, we can express the normalized area for region 2 as

$$a_{\theta,R2} = \sqrt{160\theta'^2 - 3} + \frac{8}{3\sqrt{3}}\left(\pi - 6\cos^{-1}\frac{\sqrt{3}}{4\theta'}\right)\theta'^2$$

$$\text{For } \frac{\sqrt{3}}{4} \le \theta' \le 1 \tag{5.20}$$

Equation (5.20) includes three terms: the first corresponding to area A (and C) and the second and third terms representing area C. Checking for the boundary conditions at the interface with region 1, it can be seen that the area associated with area A is zero, and thus the first and last terms are also zero because the area is a full circular area without the deductions of triangular elements of area A. In fact, under this condition, (5.19) simplifies to (5.15), which is the interface with region 1.

When $\theta = \theta_{eoc}/2$, the segment relating to area B disappears, and the second and third terms equate and cancel out. The first term becomes equal to unity as expected since it covers the cell area which when normalized area is unity.

Equation (5.20) may be unwieldy to work with and a polynomial approximation has been found as follows. However, it will be seen that even a straight-line approximation is adequate at system level.

$$a_{\theta,R2} = -4.7612 + 23.299\theta' - 23.558\theta'^2$$

$$\text{For } \frac{\sqrt{3}}{4} \le \theta' \le 1 \tag{5.21}$$

Using (5.14) and (5.20), the complete expression for the normalized area over both region 1 and 2

$$a_{\theta,R1} = \frac{8\pi}{3\sqrt{3}}\theta'^2$$

$$\text{For } 0 \le \theta' \le \frac{\sqrt{3}}{4}$$

$$a_{\theta,R2} = \sqrt{160\theta'^2 - 3} + \frac{8}{3\sqrt{3}}\left(\pi - 6\cos^{-1}\frac{\sqrt{3}}{4\theta'}\right)\theta'^2 \tag{5.22}$$

$$\text{For } \frac{\sqrt{3}}{4} \le \theta' \le 1$$

Equation (5.22) is displayed in Figure 5.5 as a function of θ' and θ'^2. Notice that in the latter case, the function is linear from the origin to a point (3/16, 0.907). This is particularly interesting when we look at the EIRP as a function of cell area as illustrated in Figure 5.6.

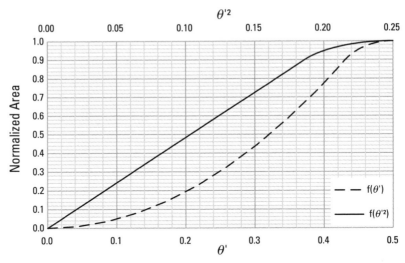

Figure 5.5 Plot of the area normalized to the hexagonal cell as a function of θ' and θ'^2.

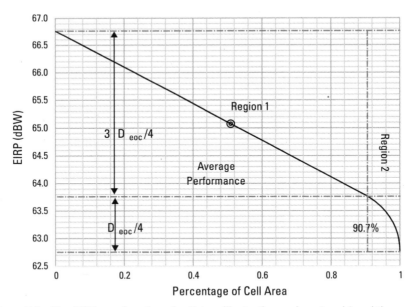

Figure 5.6 The EIRP as a function of cell area illustrating regions 1 and 2 and the associated characteristics.

The EIRP plot is produced by using the θ' as a common variable to calculate the normalized area and the EIRP. This is mainly because of region 2 as for region 1 an elegant solution is given in (5.16). A peak EIRP of 66.8 dBW is assumed with the eoc being 4 dB down, thus defining $\Delta D_{eoc} = 4$ dB. The two regions are clearly visible with the boundary condition given when the normalized area is 90.7%. In region 1 the EIRP decreases linearly from the peak to 63.8 dBW (i.e., 3 dB down or $3\Delta D_{eoc}/4$). Region 2 follows the curve given in (5.20) [or its approximation in (5.21)]. The drop in EIRP is 1 dB (i.e., another 1 dB down or $\Delta D_{eoc}/4$). It should be noted that region 2 occupies a relatively small part of the curve, and consequently if assume a linear interpolation between the start and the end points, the error in this case is about 0.3 dB. We can express the EIRP over the hexagonal cell for both regions using the straight-line approximation for region 2 as

$$\text{EIRP}_\theta = \text{EIRP}_{Pk} - \Delta D_{eoc}\frac{3\sqrt{3}}{2\pi}a_{\theta,R1}$$

For $0 \le a_{\theta,R1} \le a_H$

$$\text{EIRP}_\theta \approx \text{EIRP}_{Pk} - \frac{\Delta D_{eoc}\left(3 - 4a_H + a_{\theta,R2}\right)}{4\left(1 - a_H\right)}$$

For $a_H \le a_{\theta,R1} \le 1$

(5.23)

The weighted average EIRP over the area is easy to estimate since it can be expressed as the weighted averages for the two regions. Using the straight-line approximation for region 2, the expression becomes

$$\overline{\text{EIRP}} = \text{EIRP}_{Pk} - \frac{3\Delta D_{eoc}}{8}\frac{\pi}{2\sqrt{3}} - \frac{7\Delta D_{eoc}}{8}\left(1 - \frac{\pi}{2\sqrt{3}}\right)$$

$$= \text{EIRP}_{Pk} - \left(\frac{7}{8} - \frac{\pi}{4\sqrt{3}}\right)\Delta D_{eoc}$$

(5.24)

For the example considered, (5.24) gives an average EIRP of 65.1 dBW. A numerical integration was also performed giving the same value showing that the straight-line approximation for region 2 is good enough.

Putting (5.24) in (5.16), the area can be estimated to be

$$a = \frac{7\pi}{12\sqrt{3}} - \frac{\pi^2}{18}$$

(5.25)

This equates to 51% of the area from the center of the cell. This is also displayed in Figure 5.6.

Starting from a simple EIRP expression, we have managed to derive a simple characterization of the EIRP profile over a hexagonal cell, which is the useful part of the beam and an elemental component of the service area.

5.4   The Forward Downlink Interference Model

In conventional payloads the intrasystem interference is quite limited. In HTS systems the intrasystem is considerable and therefore merits study. This section examines the structure of the mechanism whereby intrasystem interference is generated. In a large system, this is complex, but we produce a first-order model for the copolar cells that provides us with an understanding. The section also discusses cross-polar cell interference together with the possible gateway contributions that add up to the interference scenario.

The forward uplink is the link from the gateway to the satellite, and this is considered to be more classical. However, the downlink involved a high order of frequency reuse and therefore the ensuing intrasystem interference merits specific analysis. Following our discussion in Chapter 4 about the color schemes for the frequency plan and the antenna system, Figure 5.7 illustrates the first-order interference scenario for a four-color scheme where a cell of a given color is surrounded by other cells of the same color. The same color scheme is assumed for both the frequency plan and the antenna system. The other neighboring cells are orthogonal in frequency and/or polarization. Thus, only the same color cells are of interest. For simplicity, the analysis is limited to the first set of cells closest to the cell of interest. This is the reason for referring to it as the first order. It can also be assumed that the interference scenario improves as the order of the set cells increases; the total number of cells increase but the sidelobe level drops faster.

In the first-order set there are six cells. The main issue in the forward link is the downlink where the terminal will receive the wanted signal plus six other interfering signals within the same band from the six neighboring cells shown in Figure 5.7. Actually, Figure 5.7 shows the complementary situation where the intended beam to a cell is causing interference to the six other cells.

The signal being received by a terminal can expressed as the wanted signal, C_w and the set of interfering signals, C_{i1}, C_{i2}, C_{i3}, C_{i4}, C_{i5} and C_{i6}. The separation between the centers of the wanted cell and an interfering cell is $\sqrt{3}\theta_{eoc}$ (as θ_{eoc} is the full or double-sided angular beamwidth) so that the worst

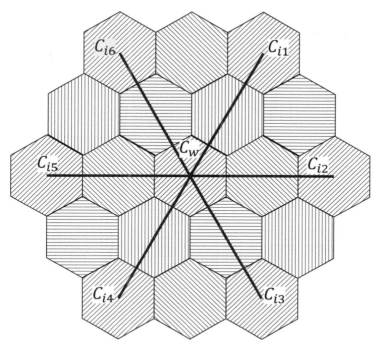

Figure 5.7 The interference scenario for a four-color scheme for the first order set same color scheme.

case for a single-entry interference is when the terminal is at the eoc where the boresight of the closest interfering beam is at a distance of $3\sqrt{3}\theta_{eoc}/4$. In the downlink interference scenario, the sources (or boresights) of the interfering signals are fixed at the spacecraft. It is the location of the terminal within the cell surface that changes together with the relative distance between the terminal and the boresights of the wanted and interference beams. Within a center-to-center cut, the wanted region of a given beam spans over the interval $(-\sqrt{3}\theta_{eoc}/4, \sqrt{3}\theta_{eoc}/4)$ while the interference region is the interval $(3\sqrt{3}\theta_{eoc}/4, 5\sqrt{3}\theta_{eoc}/4)$. This is illustrated in Figure 5.8 where part of the main lobe is used for the cell coverage but the sidelobes fall on the neighboring cell and cause copolar interference for the same color cells. Although this may be a clear illustration, the situation is more complex over a surface.

The communication traffic in each cell is considered to be uncorrelated so that the contributing fields of the interference beams are aggregated on a

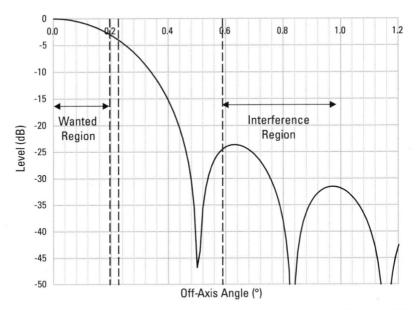

Figure 5.8   The directivity for a center-feed antenna with a diameter of 2.6m at 19.7 GHz for a parabolic-on-pedestal illumination for $n = 1$ and for an edge taper of 15 dB showing the coverage over the wanted region and the interference region.

power basis. Thus, from the antenna point, the aggregate interference level within the wanted cell and limiting the analysis to the first-order cells is given as

$$\text{EIRP}_{\text{IT}} = \sum_{m=1}^{6} \text{EIRP}_{\text{Im}}\left(u_m\right) \qquad (5.26)$$

where $\text{EIRM}_{\text{Im}} = P_{\text{TWT},M} - L_{op,m} - L_{\text{ant},m} + D_m(u_m)$ and $\text{EIRM}_{\text{Im}}(u_m)$ is the interference contribution of the mth beam at location u_m.

The EIRP equation is the same form as (5.7), but in a real system, the m paths are different, and therefore the values of the parameters of each path are also different. The simplified expression for the antenna directivity of (5.6) is not valid beyond the main lobe of the antenna and (5.2) and (5.3) could be used in a theoretical case.

Thus, (5.26) is effectively the summation of the six patterns from the interference contributions of the neighboring cells onto the wanted cell surface. In a simplified model, we assume that the repeater output power to each antenna port is identical and that the antenna gain patterns with respect to their respective boresights are also identical (i.e., that each parameter for each path is equal).

The interference level is given by the sum of the EIRP patterns over the wanted cell surface. In the simplified model, the resulting pattern is displayed in Figure 5.9 for the antenna system discussed above. The six-entry interference scenario produces the symmetry in the pattern. The first lobe in Figure 5.8 can be observed as the rounded rim due to the combined effect of the six interference beams. The null effect is visible as being perfectly central because of the assumption that the interfering beams are completely axisymmetric and identical. In a real system this perfection will not exist.

However, this is not really the C/I but a representation of the aggregate interference level. The wanted carrier needs to be considered for the C/I estimation. Thus, for a given terminal location within the wanted cell C_w, the carrier-to-aggregate interference for the forward downlink can be expressed as

$$\frac{C}{I}_{\text{FWD,DL,}} = \frac{\text{EIRP}_W\left(u_W\right)}{\sum_{m=1}^{6}\text{EIRP}_{\text{Im}}\left(u_m\right)} \tag{5.27}$$

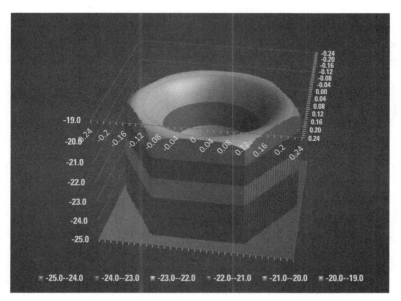

Figure 5.9 Illustration of the intrasystem interference on the wanted cell from the first-order same color interfering cells from a theoretical antenna system with a diameter of 2.6m at 19.7 GHz for a parabolic-on-pedestal illumination for $n = 1$ and for an edge taper of 15 dB.

This is illustrated Figure 5.10 where is C/I is given within the wanted cell. The rounded rim has now disappeared, and instead there is a crown like structure due to the slope of the carrier from the wanted beam.

Of course, in a real antenna system with a multifeed and an off-set configuration, the symmetry will be reduced. Asymmetry will be introduced in the side-lobe structure of each beam around the axis when taking into account that feeds will be on a surface around the focus point and obviously not all at the focus point. These effects result in defocus aberration. There are also multipath effects from the spacecraft wall that can distort the pattern. The low level of the sidelobes makes them particularly sensitive to such impairments. The well-like structure in the center of the pattern in Figure 5.9 or Figure 5.10 will be flattened and shallower due to these impairments. Additionally, each beam will experience a slight difference in level that further degrades the symmetry.

The discussion thus far has concentrated on copolar interference. However, examination of Figure 5.7 shows that there are two cells adjacent to the wanted cell that are cofrequency and cross-polar, and that will contribute a cross-polar interference in the wanted cell. The center-to-center spacing is

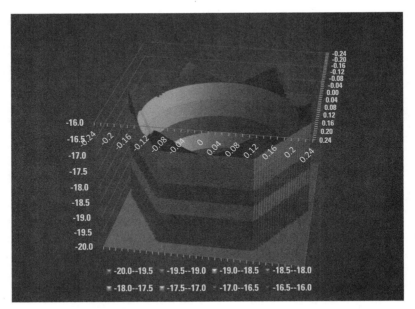

Figure 5.10　Illustration of the intrasystem C/I within the wanted cell from the first-order same color interfering cells from a theoretical antenna system with a diameter of 2.6m at 19.7 GHz for a parabolic-on-pedestal illumination for $n = 1$ and for an edge taper of 15 dB.

$\sqrt{3}\theta_{eoc}/2$, but there is no separation at the border edge. There are two such adjacent cells that are on diametrically opposite sides of the cell. This means that although there are two contributions in the simplified model, the significant net contribution at a given location is predominantly from a single contribution of the closer cell.

The cross-polar performance of off-set antennas is a function of the focal length to diameter ratio (*F/D*) and tends to improve for a higher *F/D*. For a spacecraft reflector with a diameter of about 2.6m or less, a *F/D* of one or more is typical. In this range of *F/D*, the cross-polar component is typically lower than the sidelobe level in the region of interest, and therefore the contribution to the intrasystem interference is small or negligible. However, as the reflector diameter increases, it will be increasingly difficult to maintain a *F/D* in this range. Therefore, the cross-polar performance will degrade and the cross-polar contribution to the intrasystem interference becomes more significant.

Besides the downlink of the forward link, which is the most significant, the intrasystem interference between the gateways of the uplink is also another contribution. The objective would be to maximize the antenna spatial isolation between the set of sites of the gateways within the designated area, and this typically translates into maximizing the distance between the sites. The gateway tends to be a large station and therefore possesses a very narrow beam. However, all gateway antennas are pointed at the same satellite so it is mainly the isolation at the spacecraft level that determines the system isolation performance.

Considering an HTS system over a given region, as the system gets larger, the number of gateways within the region increases, so that even before getting into operational issues, the intersite distance decreases and the isolation decreases.

However, the intersite isolation is not the only criterion at system level, and often there is not complete liberty in the selection of the sites. As the name suggests, the gateway sites need to have good fiber connectivity to ensure that there is the required capacity and that the required service quality can be maintained. The sites need to be accessible for maintenance, and preferably teleports should be employed for infrastructure and operational reasons.

No matter the performance and reliability of the gateways, it can be assumed that each gateway will have outages due to atmospheric propagation and for maintenance reasons. It is thus interesting to have a satellite that can support G-for-(*G+S*) gateways (i.e., *G* is the active and operational number of gateways with *S* being the standby gateways). Such a system is described in [4] where a payload can select *G* gateways out of *G+S* possibilities so that a gateway in the set of the *G* active gateways suffering an outage can be replaced by a standby gateway. Of course, there is an additional system cost involved, and ideally this functionality requires support from the network, although there may be operational workaround solutions.

This section has examined the topological structure of the generation of interference from copolar cells and then elaborated a model that is simplified to the first order. In addition, the section has discussed the results to provide an understanding of the mechanism and described the mechanism of the generation of interference from the cross-polar cell and gateway when applicable.

5.5 Other Impairments

The intrasystem interference from the adjacent cells is not the only impairment to the link budget. This section addresses other possible sources of link budget degradation such as pointing accuracy and handover of cells in mobility.

With pencil beams, antenna pointing becomes an important issue. Gateways with apertures of 6m in diameter or more operating in the Ka-band or above make an antenna-pointing system a must. These systems optimize the signal strength toward the spacecraft but in an indirect fashion. The spacecraft provides a stable beacon that is used to control the antenna pointing to maximize the receive signal. Correlation between the transmit and receive beams is assumed.

The spacecraft beam pointing accuracy is a function of two effects: platform pointing and antenna distortion. Platform pointing depends on the precision of the attitude control and on the platform thermal distortion. The attitude control maintains the spacecraft attitude fixed relative to a given point on the platform, typically on the Earth-facing deck of the spacecraft. The feed systems are also mounted on the Earth-facing deck. They are protected by their own thermal system to minimize the temperature excursions.

The fixture points of the reflector's are often a few meters away from the attitude reference point. The reflectors suffer from thermal distortion over the operating temperature range. Specific designs and precautions are taken to minimize the thermal excursions. Additionally, the reflectors can be designed to be desensitized over a temperature range. The net effect is a small beam distortion and shift. Beacons can also be used to minimize this effect. The beacon is emitted from Earth, and the spacecraft antenna system includes a RF-sensing closed-loop system that uses the information to steer the reflector to the optimal position. The system is dedicated to a reflector so that with a four-antenna spacecraft system, the four antennas require individual control.

It should be noted that there is some correlation between the distortion suffered by the beams originating from the same reflector. This helps to maintain the same color interference performance assuming that the same color beams are associated to a given reflector. In an open-loop system, there is also some correlation between beams generated from the reflectors from the same side of the spacecraft. The shift of the performance over a cell implies an increase in the performance on one side at the cost of the diametric side, so the overall impact on capacity may be minimal. However, there may be an impact on the availability in that part of the cell toward the edge where the performance is degraded. This effect tends to have a diurnal periodicity.

When considering an HTS for mobility, as a terminal is moving on a trajectory, several cells may be crossed. The network is designed for beam switchover, transferring traffic from one cell to another. In this case, moving terminals and moving beams have a similar effect, and closed-loop spacecraft antenna pointing may not be required especially when associated with dynamic cell definition.

The signal being processed by each TWTA in the payload is typically composed of a number of wideband streams. The nonlinearity effects are often quantified in terms of the noise-power ratio (NPR). As usual, the operating point of the TWTA is a compromise between the available RF power and the linearity requirements. A linearized TWTA ensures the lowest back-off for a given NPR. Typically, the operating point for the forward link is between 2 and 3 dB to deliver an NPR between 15 and 17 dB.

There are other impairments from the repeater. The filter performance is not completely flat. Since the filters are wideband, inband ripple may become more of an issue. These effects can can be minimized by design of the payload. Additionally, precompensation on the uplink signal at the gateway can be used as a mitigation technique. There are also potential multipath issues, especially when a TWTA is used to drive two beams [5]. However, the effect can be reduced as discussed in Chapter 4.

5.6 ACM

Since ACM significantly contributes to the economic success of HTS systems, it is useful to understand how the HTS system benefits from it and its performance.

ACM enables link budgets to be optimized for a given location in the service area under the prevailing climatic conditions. The traditional approach had been to determine the link budget requirements for the clear-sky conditions and then add margin to ensure that the required availability was supported over the service area.

This is rather wasteful, however, since it is impossible to sustain a constant margin for a given availability over a given service area. The implication is that the management of meeting the commercial requirements is complicated since several terminal sizes had to be used. More importantly, for most of the time, the margin is not required and not put to commercial use. Thus, for example, for an availability of 99.8%, the system is overdimensioned for 99.8% when it is dimensioned for 0.2% of the time, and this is for the critical case within the service area. The higher the availability of the system, the more the system is overdimensioned for a higher portion of the time and for a smaller section of the service area. Additionally, since the number of terminal sizes has to be limited commercially, most of the system is overdimensioned for a portion of the time that exceeds the availability figure.

All this shows that a system can be tweaked to maximize capacity and become more economically advantageous by using margins in a more judicial fashion. This is the dimension of the envelope in which ACM pushes: minimizing dead margins while maximizing capacity for the maximum portion of the time, rendering a system more profitable while meeting the service-level agreement (SLA).

ACM was introduced in the DVB-S2 standard, which was first issued by the European Telecommunications Standards Institute (ETSI) in 2004 [6]. It is also part of the subsequent standard DVB-S2x [7].

The ACM specification is normally expressed in the form of a modulation and coding table with the associated E_s/N_0 and spectral efficiency. The combination of a modulation scheme and a coding scheme is referred to as a MODCOD. The modulation and coding define the spectral efficiency, which can be defined in bits per symbol or bits per second per hertz.

$$\frac{\text{bits}}{\text{symbols}} = \frac{R_b}{R_S} \tag{5.28}$$

where R_b is the information bit rate, and R_S is the symbol rate.

Now the symbol rate is related to the required bandwidth by the roll-off so that

$$B = \left(1 + \alpha_{ro}\right)R_S \tag{5.29}$$

where α_{ro} is the roll-off, and $\alpha_{ro} < 1$.

In DVB-S2, there are two concatenations of an outer code and an inner code, the Bose-Chaudhuri-Hocquenghem (BCH) multiple error correction binary block code and the low-density parity check (LDPC) codes, respectively. The BCH codes employed can correct eight to 12 errors per block. The coding frames are 64,800 bits or 16,200-bits-long. Besides the concatenated coding, bit interleaving is also employed where bits are interleaved between three, four, or five frames for 8PSK, 16APSK, and 32APSK, respectively. Interleaving randomizes bursts of errors while concatenation mitigates the error floor effect of the LDPC.

In DVB-S2x, the concatenation structure of BCH and LDPC is retained although additional coding schemes are introduced. The depth of the interleaving is further extended for the higher-order modulation. The lengths of coding frames of DVB-S2 have also been maintained but another length of 32,400 is introduced for very long signal-to-noise ratio (VL-SNR) operation.

The ultimate performance in channel capacity is the Shannon limit [8], which is expressed as

$$C_{cap} = B\log_2(1 + S/N) \tag{5.30}$$

where:

C_{cap} is the capacity;
B is the bandwidth;
S/N is the SNR.

Alternatively, we can write the spectral efficiency in bits per second per hertz as

$$\eta_{spec} = \log_2(1 + S/N) \tag{5.31}$$

Figure 5.11 illustrates the MODCOD performance of DVB-S2x together with Shannon's limit. In DVB, the spectral efficiency is given in bits per symbol, but the relationship of symbol rate and bandwidth is given in (5.29).

Table 5.3 provides some of the salient features of the ACM of DVB-S2 and DVB-S2x, showing the dynamic range, the granularity, and the performance with respect to the Shannon limit. For comparison, α_{ro} is assumed to be zero, which adds some optimism. A low granularity reduces the system overhead.

This section has considered ACM to assess its system advantages, examining the ACM as given in the latest standard, DVB-S2x, and reviewing its performance in absolute terms and relative to the previous standard DVB-S2. We have also deduced some top-level characteristics that are useful at system level.

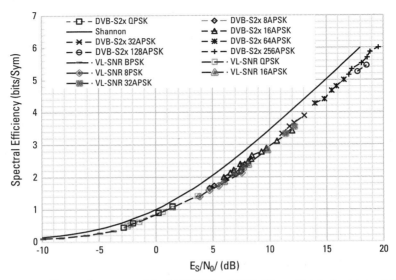

Figure 5.11 The performance of the MODCODs of DVB-S2x, which is the basis of the ACM function together with Shannon's limit as given in (5.31).

Table 5.3
A Comparative Table of the Salient Features of the ACM for DVB-S2 and DVB-S2x

	DVB-S2	**DVB-S2x**
Dynamic range of Ideal E_s/N_0 (dB)	−2.35–16.05 dB[1]	−2.85–19.57 dB[1]
	—	−9.9–19.57 dB[2]
Granularity over the dynamic range	0.68 dB[‡]	0.60 dB[3]
Average discrepancy with respect to Shannon	2.07 dB[‡]	1.51 dB[3]
	2.07 dB[§]	1.50 dB[4]
	—	4.02 dB[5]
	—	3.91 dB[6]

[1] For an ideal channel and a coding frame of 64,800 bits.

[2] Employing VL-SNR mode.

[3] Over the full available dynamic range with an ideal channel and a coding frame of 64,800 bits.

[4] Over the available dynamic range limited to $E_s/N_0 = 16$ dB with an ideal channel and a coding frame of 64,800 bits.

[5] Over the full available dynamic range with a nonlinear hard-limiter channel and a coding frame of 64,800 bits.

[6] Over the available dynamic range limited to $E_s/N_0 = 16$ dB with a nonlinear hard-limiter channel and a coding frame of 64,800 bits.

5.7 The Forward Link Budget

The system level performance is often summarized in the system capacity, which typically includes contributions from both the forward and the return. Sections 5.1–5.6 discuss the elements we need to develop a system link budget analysis for the forward link.

For the forward link, the link budget performance is basically dominated by downlink performance. The uplink from the gateway means that the uplink can be dimensioned such that its performance impact on the total link budget is small or negligible. Table 5.4 gives the reference link budget under clear sky conditions at the beam boresight. It also provides the underlying assumptions for the ground segment. The gateway is assumed to have a RF TWTA of 500W with about 12-dB ULPC. The spacecraft EIRP and antenna performance is as previously discussed and illustrated in Figures 5.6 and 5.8.

Elements of the reference link can be used to estimate the performance over the cell. The $\left(\dfrac{C}{N+I}\right)_{\text{FWD}}$ that the terminal receives can thus be expressed as

$$\left(\frac{C}{N+I}\right)_{\text{FWD}} = -10\log_{10}\left(\frac{10^{\frac{-(C/N)_{\text{FWD,UL}}}{10}} + 10^{\frac{-(C/N)_{\text{Rptr}}}{10}} + 10^{\frac{-(C/I)_{\text{FWD,RPTR}}}{10}} + 10^{\frac{-(C/I)_{\text{FWD,RPTR}}}{10}}}{+ 10^{\frac{-(C/N)_{\text{FWD,DL}}}{10}} + 10^{\frac{-(C/I_C)_{\text{FWD,DL}}}{10}} + 10^{\frac{-(C/I_x)_{\text{FWD,DL}}}{10}}}\right)$$

$$(5.32)$$

where:

> $(C/N)_{\text{FWD,UL}}$ is the uplink C/N in decibels and is fixed for a given gateway;
>
> $(C/N)_{\text{Rptr}}$ is the repeater impairment in decibels, which is driven by the NPR that is applicable for a given TWTA;
>
> $(C/N)_{\text{FWD,DL}}$ is the downlink carrier-to-noise ratio in decibels, which is a function of the radial distance of the terminal from the beam boresight or cell center;
>
> $(C/I)_{\text{FWD,RPTR}}$ considers the impairments from the repeater due to the return effect;
>
> $(C/I_C)_{\text{FWD,DL}}$ is the downlink co-frequency copolar C/I in decibels from the six-cell cluster in the model under discussion, which is a function of the terminal location within the cell as illustrated in Figure 5.10;

Table 5.4

Reference Clear-Sky Link Budget for the Forward Link at the Beam Boresight

Uplink			
Earth Station		**Link**	
Frequency	28.00 GHz	Uplink EIRP	73.6 dBW
HPA	30.0W	Pointing accuracy	0.05°
Output losses	3.00 dB	Pointing loss	1.10 dB
OBO	4.90 dB	Range	37,236 km
Antenna diameter	9.10m	Free-space loss	212.82 dB
Antenna beamwidth	0.08°	Spreading loss	162.41 dB
Antenna gain	66.7 dBi	Satellite G/T	20.0 dB/K
Uplink EIRP	73.6 dBW	Input flux density	-89.9 dBW/m^2
		C/N_0	108.3 dBHz
Downlink			
Earth Station		**Link**	
Frequency	19.70 GHz	Satellite EIRP	66.8 dBW
Antenna diameter	0.70m	Range	37,236 km
Antenna beamwidth	1.52°	Free-space loss	209.77 dB
Antenna gain	41.4 dBi	Earth station G/T	17.2 dB/K
Antenna temperature	40.0K	C/N_0	102.7 dBHz
Receiver noise factor	2.5		
Receiver noise temperature	225.7K		
System temperature	265.7K		
	24.2 dBK		
G/T	17.2 dB/K		
Total		C/N_0	101.7 dBHz
		Bandwidth	1000 MHz
		C/N	11.7 dB

$(C/I_X)_{FWD,DL}$ is the downlink co-frequency cross-polar C/I in decibels from the two adjacent cells, which is a function of the terminal location.

The exercise has been performed for the defined system assuming that the satellite TWTA has an available bandwidth of 500 MHz, 750 MHz, 1,000 MHz, 1,250 MHz, and 1,500 MHz. The synthesis of the performance is given in Figure 5.12. In all cases, the resources in terms of the antenna and TWTA are the same. The TWTA is assumed to be operated with a 2.4-dB output back-off and an associated NPR of 16 dB. DVB-S2x is employed in all cases with a 2.5-dB allowance for the implementation of the ACM and other effects. The cross-polar C/I is assumed to be single entry at 25 dB.

As can be expected, the mix of modulation used changes. At the low bandwidth side, 16APSK is more extensively used with no use of QPSK under clear-sky conditions. At 1,000 MHz, the split is more balanced, while at the high bandwidth end there is a significant shift toward QPSK. It can be expected that the system availability is better at 500 MHz than at 1,500 MHz.

The average spectral efficiency over the cell decreases with increasing system bandwidth. To a first order in this example, the C/N decreases inversely proportionally to the bandwidth, and the spectral efficiency decreases with

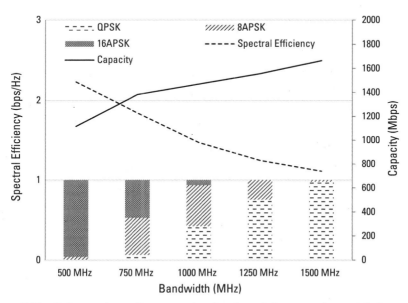

Figure 5.12 Cell capacity performance showing the split between the modulation schemes and the capacity and spectral efficiency.

a log function with respect to the C/N as given in (5.31). The cell capacity increases with the system bandwidth. To a first order, this follows (5.30). The capacity is a function of the bandwidth and the C/N. The capacity is a log function of the C/N, which drops with bandwidth. However, the capacity increases linearly with bandwidth. Since the linear function is stronger than the log function, the capacity increases with bandwidth despite the reduction in the C/N. This is, of course, just one cell, and the analysis can be produced for each cell to produce the system capacity.

Sections 5.1–5.6 discussed the various contributions to the capacity assessment for the forward link. This section concluded the chapter by developing a link budget analysis for the forward link that can be extended over the service area to estimate the forward capacity. Chapter 6 investigates the return link capacity.

References

[1]   Fenech, H., et al., "High Throughput Satellite Systems: An Analytical Approach," *IEEE Transactions on Aerospace and Electronic Systems*, Vol. 51, No. 1, January 2015, pp. 192–202.

[2]   Lo, Y.T., and S. W. Lee, *Antenna Handbook: Volume II, Antenna Theory*, New York: Van Nostrand Reinhold, 1993.

[3]   Stutzman, W. L., and G. A. Thiele, *Antenna Theory and Design*, John Wiley & Sons, Inc., 1998.

[4]   Fenech, H., and E. Lance, *Telecommunications Network*, European publication: EP2104243 (B1), granted on May 4, 2011; French publication FR2929059(B1), granted on October 26, 2011; U.S. publication US8149869 (B2), granted on April 3, 2012.

[5]   Fenech, H., and E. Lance, *Payload for Multi-Beam Satellite*, French patent: FR2950496 (B1), granted October 21, 2011; U.S. patent US9118384 (B2), granted August 25, 2015.

[6]   *Digital Video Broadcasting (DVB); Second Generation Framing Structure, Channel Coding and Modulation Systems for Broadcasting, Interactive Services, News Gathering and other Broadband Satellite Applications; Part 1: DVB-S2*, ETSI 302 307-2 V1.1.1 (2014-10), European Telecommunications Standards Institute (ETSI), Sophia Antipolis, France.

[7]   *Digital Video Broadcasting (DVB), Second Generation Framing Structure, Channel Coding and Modulation Systems for Broadcasting, Interactive Services, News Gathering And Other Broadcasting Satellite Applications, Part 2: DVB-S2 Extensions (DVB-S2X)*, ETSI 302 307-2 V1.1.1 (2014-10), European Telecommunications Standards Institute (ETSI), Sophia Antipolis, France.

[8]   Shannon, C. E., "A Mathematical Theory of Communication," *The Bell System Technical Journal*, Vol. 27, No. 3, July 1948, pp. 379–423.

6

HTS System Analysis: The Return Link

6.1 Overview

This chapter addresses the return link and so can be considered as the sister chapter to Chapter 5, which covers the forward link.

First, Section 6.2 looks at the antenna. Having dimensioned the antenna configuration for the forward link, we now consider the implications for the return link.

In Section 6.3, a satellite G/T model for the user link is developed in a similar fashion to the development of the EIRP model in Chapter 5.

Next, Section 6.4 analyzes the user uplink interference scenario, developing two models: a model where the interference antenna pattern is approximated to a straight line and another in which the curvature of the contour is considered. Section 6.4 also presents a technique for modeling power control.

Section 6.5 examines the DVB-RCS specification, and finally, Section 6.6 describes a link budget that looks at the effect of the available bandwidth.

While this analysis is theoretical, it should provide a system-level overview with insight into the key system elements.

6.2 The Antenna

Chapter 5 details the fundamentals of the antenna. This section focuses on the aspects of the antenna that are pertinent to the return link and addresses the case in which the reflector is shared between the forward and the return.

The antenna is a reciprocal device: An antenna that transmits at a given frequency and a given polarization with a certain radiation pattern will have the same pattern for reception at the same frequency and polarization. In practice, the design of an antenna that includes the transmit function would have to consider power handling.

In HTSs, single-beam-per-feed (SBPF) antenna techniques are often used; this implies that the same feed is used for both uplink and downlink functions. However, the uplink and the downlink employ different frequency bands and are frequently orthogonal in polarizations. This facilitates the terminal antenna design. In the Ka-band, the uplink band is at around 29.5 GHz, and the downlink is at around 19.7 GHz. Normally, circular polarization is utilized.

At first order the performance of the antenna will scale with frequency. The physical dimensions of the feed and the reflector are obviously the same, but the electrical dimensions are different. The reflector with a physical diameter of 2.6m has an electrical diameter of 170λ and 256λ at 19.7 GHz and 29.5 GHz, respectively, so that the antenna becomes more directional at the uplink band. The directivity increases, and the beamwidth decreases. Figure 6.1 illustrates the scaling due to frequency.

The increased directivity at boresight is a welcome feature since it facilitates the G/T, but this indirectly presents challenges to the cell coverage. For the network management purposes, the cell definition is identical for both the uplink and the downlink, but the increased directivity on the uplink implies that the directivity variation over the cell is larger than that on the downlink. The difference of the two frequencies discussed is a factor of 1.5 so that the beamwidth at a relative level for the uplink is a factor of 1.5 less than that of the downlink at the same level.

In the forward downlink case presented in Chapter 5, the eoc directivity performance was 4 dB below the peak, and this was determined to be at a beam diameter of about 0.45°. At the uplink frequencies, the same level occurs at a diameter of 0.3°, a factor of 1.5 less. To attain the required beamwidth for the return, the contour at 9.9 dB would have to be used. This can be seen in Figure 6.1.

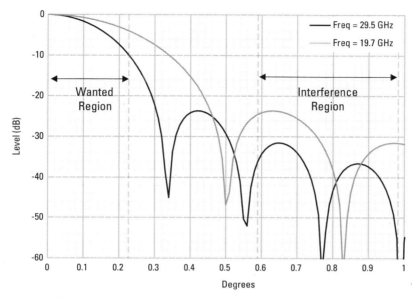

Figure 6.1 The performance for a center-feed antenna with a diameter of 2.6m at 29.5 GHz and 19.7 GHz (for a parabolic-on-pedestal illumination for $n = 1$ and for an edge taper of 15 dB) showing the coverage over the wanted region and the interference region.

The situation is a bit more complicated since the physical dimension of the feed is also fixed for both the uplink and the downlink. The forward link capacity tends to be more valuable in an HTS system so that the feed is mostly optimized for the downlink: The gains on directivity reduce the amplification requirements, which is a driving parameter in the power and mass equation of spacecraft launch. Thus, the illumination of the reflector by the feed for the uplink is less optimized. The frequency factor of 1.5 is quite large, and the implication is that the illumination of the feed at the uplink frequency may extend beyond the first null of the feed. This translates to increasing the edge taper and effectively using a smaller portion of the reflector, which causes the beam to broaden.

Normally the Ka-band HTS systems utilize circular polarization. This leads to beam squint where the peak of the beam shifts to one side of the symmetry plane of an offset reflector antenna, and the peak of the other polarization shifts to the other side of the symmetry plan. This effect increases with the offset angle and θ_{3dB}, and decreases with the F/D. The shift of the beam peak with respect to the antenna boresight can be approximated as follows [1–3]:

$$\frac{\theta_s}{\theta_{3dB}} = \frac{\sin(\theta_0)}{4\pi F/D} \tag{6.1}$$

where θ_s is the angular offset of the beam peak with respect to the antenna boresight, and θ_0 is the angle between the incident beam and the reflected beam from the main reflector in the plan of symmetry.

In an HTS case considered above, the squint is not symmetric since the orthogonal beams considered have considerably different frequencies. What matters is the sum of the beam squints, since this must be managed at the cell-definition level. It so happens that for the configuration being considered, the difference between the uplink and downlink peaks is small and on the order of 0.02°, which can be managed at the system level. However, as antenna diameters increase, θ_0 increases, and as the F/D decreases due to accommodation issues, it becomes more significant.

6.3   The User G/T Model

In this section we use the G/T expression together with the simple antenna directivity relationship from Chapter 5 to derive a user G/T model. As in Chapter 5, a model is developed to give the profile of the G/T over a theoretical hexagonal cell. Since the geometry analysis over a cell is identical to that of the EIRP model given in Section 5.3, maximum reuse will be made of the results already produced.

The development of the return user G/T model is similar to that of the forward EIRP. The G/T in a given direction is given as

$$\frac{G}{T_{\theta,\varphi}} = D_{\theta,\varphi} - L_{\text{UL.ant}} - T_{\text{UL.ant}} - T_{\text{Rptr}} \tag{6.2}$$

where:

$D_{\theta,\varphi}$ is the antenna directivity of the given beam in the given direction in dBi;

$L_{\text{UL.ant}}$ represents the uplink antenna losses up to the antenna-repeater interface in decibels;

$T_{\text{UL.ant}}$ is the noise temperature of the uplink antenna referred to the antenna-repeater interface in dBK;

T_{Rptr} is the repeater noise temperature referred to the antenna-repeater interface in dBK.

The uplink antenna temperature on the spacecraft is a function of the brightness temperature that the satellite is looking at, along with the antenna design. The uplink antenna is looking at the Earth, and therefore the Earth brightness temperature will be the input to the estimation. Earth brightness temperature models exist [4] that have to be integrated over the directivity of the beam, and this gives the antenna noise temperature of an ideal antenna. Additionally, there are also contributions to the antenna noise temperature from the losses of all antenna elements up to the antenna-repeater interface.

The repeater noise temperature is primarily dependent on the noise figure of the first active element of the reception chain, which is typically the LNA of the repeater and its operating physical temperature. Considerable effort is invested in ensuring that the LNAs are as close as possible to the antenna (to minimize losses) and that the post-LNA contributions are low. Maintaining a low LNA operating physical temperature also helps.

Assuming concentric alignment of the beam and the cell and an axisymmetric antenna, the G/T of an HTS return payload can be expressed using the simplified equation for the main lobe pattern of the beam as employed for the forward EIRP to be

$$\frac{G}{T}_{\theta} = \frac{G}{T}_{Pk} - 4\Delta D_{eoc} \left(\frac{\theta}{\theta_{eoc}} \right)^2 \tag{6.3}$$

where:

G/T_0 is the G/T in dB/K at an off-axis angle θ of a given beam;

G/T_{Pk} is the peak G/T in dB/K at the boresight of that beam and the cell since both are assumed to concentric;

θ_{eoc} is the full eoc beamwidth in degrees.

Following the same methodology as that of the forward EIRP, we can divide the hexagonal cell into two regions: regions 1 and 2. Region 1 is the surface within a circle inscribed in the hexagonal cell, while region 2 is the remaining surface outside the inscribed circle and within the hexagon.

Using the same methodology as in Section 5.3 for the EIRP for the forward payload and employing the straight-line approximation for region 2, we can write the G/T as a function of the normalized area over the hexagonal cell as

$$\frac{G}{T}_\theta = \frac{G}{T}_{Pk} - \Delta D_{eoc} \frac{3\sqrt{3}}{2\pi} a_{\theta,R1}$$

$$\text{for } 0 \leq a_{\theta,R1} \leq a_H$$

$$\frac{G}{T}_\theta \approx \frac{G}{T}_{Pk} - \frac{\Delta D_{eoc}\left(3 - 4a_H + a_{\theta,R2}\right)}{4\left(1 - a_H\right)} \tag{6.4}$$

$$\text{for } a_H \leq a_{\theta,R1} \leq 1$$

Recall that the normalized area is normalized with respect to the area of the hexagonal cell and that the maximum normalized area in region 1 is $a_H = \pi/2\sqrt{3} = 90.7\%$.

ΔD_{eoc} is the relative contour that corresponds to the eoc level. At a_H, the G/T would be down by $\frac{3}{4}\Delta D_{eoc}$. As discussed earlier for a SFPB system, to a large extent the choice is made for the downlink. In our example, 4 dB was selected for the forward downlink, and assuming a transmit/receive function, this translates to 9.9 dB for the return uplink.

In region 2, the geometry is a bit more involved, and the normalized area can be estimated as in Section 5.3. The G/T in terms of the normalized area in region 2 can be calculated in a parametric fashion using $\theta' = \theta/\theta_{eoc}$ as the common variable for the normalized area and the G/T. However, region 2 covers less than 10% of the area, and the G/T drops a further $\Delta D_{eoc}/4$ and we simplify to a straight-line approximation.

The weighted average G/T over the area can be estimated as the weighted averages for the two regions. Using the straight-line approximation for region 2, the expression becomes

$$\frac{\overline{G}}{T} = \frac{G}{T}_{Pk} - \left(\frac{7}{8} - \frac{\pi}{4\sqrt{3}}\right)\Delta D_{eoc} \tag{6.5}$$

For the example considered and as demonstrated in Figure 6.2, the peak G/T with the given antenna at the given frequency is calculated to be 26.1 dB/K, and the eoc G/T is 9.9 dB down from the peak (i.e.,16.2 dB/K). Region 1 covers the boresight to 90.7% of the hexagonal cell area, and region 2 covers the rest of the cell. Within region 1 the G/T drops by three-quarters of the value of the relative eoc level (i.e., 7.4 dB). Thus, the boundary between the two regions occurs at (90.7%, 18.7 dB/K). The weighted average G/T is estimated to be 22.0 dB/K and is exceeded for 51% of the area.

Using the results from the antenna analysis and the cell geometry analysis performed in Chapter 5, a G/T model is produced with its distribution over the theoretical hexagonal cell.

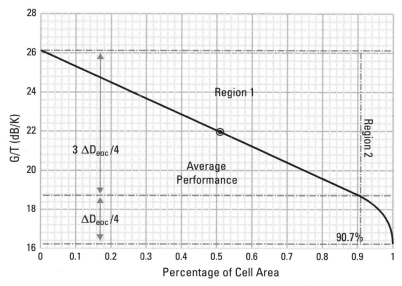

Figure 6.2 The G/T as a function of cell area illustrating regions 1 and 2 and the associated characteristics.

6.4 The Return Uplink Interference Model

Chapter 5 provided an overview of the copolar cell interference scenario for the forward. This section considers the copolar cell interference scenario for the return, which is notably different from that of the forward. This is partly because of the signal structure and partly because of the generation process.

The return downlink is the link from the satellite to the gateway, and this is considered to be more classical. However, the uplink involves a high order of frequency reuse and therefore merits specific analysis.

The return uplink interference scenario is considerably different from that of the forward downlink. The return access scheme is typically a multiple frequency-time division multiple access (MF-TDMA) (e.g., DVB-RCS), which is detailed Section 6.5. However, since we are looking at cofrequency interference, the scenario can be simplified to TDMA. The statistical occupancy of the TDMA signal could be considered, and one approach would be to condition the return traffic to the forward traffic. However, commercially we would like to exploit the return as much as possible without limiting it to the compliment of the forward capacity and other applications [e.g., satellite news gathering (SNG)] can be considered. Thus, the TDMA traffic is assumed to be such that the temporal occupancy of the TDMA is close to unity.

Similar to the forward downlink case and assuming an infinite lattice in a four-color scheme, each cell is surrounded by a first order of six cells of the same color. This situation is different here with respect to the forward downlink. For a given cell, the return uplink signal can arise from anywhere within the cell. Taking this into consideration together with the assumed signal structure, a statistical approach is required to assess the interference level as the interference signal from the population of terminals in each of the six cells arises from random locations within the respective interfering cell.

Since we are dealing with the uplink, it is also important to note that the interference is aggregated at the output port of the satellite uplink antenna. For the beam serving the wanted cell, the aggregate interference level is determined by the receive performance of the satellite antenna outside the wanted cell and on the interfering cells. This is predominately the sidelobe performance of the antenna. This means that although the interference level has a statistical nature because of the randomness of the location of the source, it is independent of the location of the wanted signal. However, the C/I is also a function of the location of the wanted signal within the wanted cell and the uplink EIRP.

The cofrequency interference is only possible between cells of the same color (i.e., cofrequency and copolar), since cells in different colors are orthogonal in frequency and/or polarization. Thus, referring to Figure 6.3, the six interfering cells are denoted as C_{i1}, C_{i2}, C_{i3}, C_{i4}, C_{i5}, and C_{i5} while the victim or wanted cell is denoted C_w. All terminals, wanted and interfering, are at random locations within their respective cell.

The first step in this analysis is to investigate the statistics of the interference within a single interfering cell. The second step is to determine the statistics of the combination of the six interferences.

Because of the higher frequency, the pattern is more compact on the return frequencies. While on the forward link, most of the first sidelobe, the second null, and parts of the second lobe were captured in the interfering cell; at the return frequencies, the second and third lobes are clearly visible together with the third and fourth null. The second and third peaks occur at about 31.5 dB and 36.6 dB down from the main peak. This is, of course, an attractive secondary feature as it means that the interfering level should be lower. Figures 6.1 and 6.4 illustrate of this effect. Figure 6.4 shows the relative directivity over an interference cell of the antenna considered for the forward link but operating in the return uplink frequency.

In the theoretical model assumed, all six cells would have identical distributions because of the complete axisymmetry. In a real system, this is rarely the case. Further, we assume a homogenous distribution of users, and initially we assume the same power for all terminals so that the relative frequency

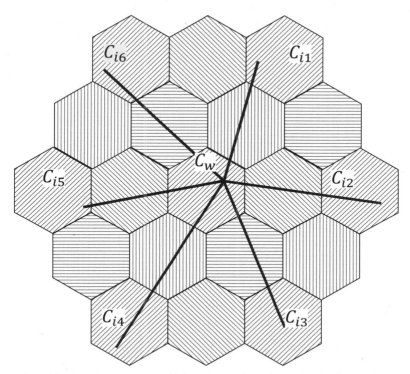

Figure 6.3 The interference scenario for a four-color scheme for the first-order set same color scheme.

distribution of the single-entry interference can be deduced from Figure 6.4. This implies that in this case there is no uplink EIRP profiling across the cells and that the uplink EIRP profile across the cell is flat.

A probability density function can be used to characterize the interference from a single cell. This can be expressed as a function of the power from all the possible locations within the cell. Since all terminals are assumed to produce an uplink with the same EIRP, the statistics can be limited to the directivity or gain, as the effective power contribution can be estimated in the link analysis. We select a cell as shown in Figure 6.5(a).

With a local coordinate system (θ, φ) for the cell as shown in Figure 6.5(b), the beam center is at $(0, -\theta_{eoc}\sqrt{3})$. The probability of a power contribution is a function of the length of the elemental contour or contours of that contour within the cell. There may be more than one length contour in a cell at a given level. We further simplified the analysis by ignoring the curvature of the contours so that the power becomes a function φ and the associated probability a function of θ.

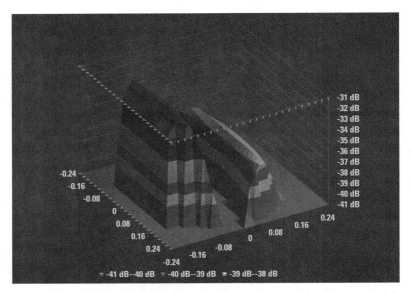

Figure 6.4 The relative directivity of a center-feed antenna with a diameter of 2.6m at 29.5 GHz for a parabolic-on-pedestal illumination for $n=1$ and for an edge taper of 15 dB over an interfering cell.

From Figure 6.5(b) and using the symmetry between the four quadrants of the hexagon and working on the right half of the hexagon, we can express θ as

$$\theta = \frac{\theta_{eoc}}{2} - \frac{|\varphi|}{\sqrt{3}}$$
$$\text{for } |\varphi| \le \frac{\theta_{eoc}\sqrt{3}}{4} \tag{6.6}$$

The probability density function (PDF) for the continuous variable Φ can thus be expressed as the length of the elemental strip in the θ-axis over half the area of the cell:

$$f_{\Phi}(\varphi) = \frac{\dfrac{\theta_{eoc}}{2} - \dfrac{\varphi}{\sqrt{3}}}{\dfrac{3\theta_{eoc}\sqrt{3}}{16}} \tag{6.7}$$
$$= \frac{8\theta_{eoc}\sqrt{3} - 16|\varphi|}{9\theta_{eoc}^2}$$

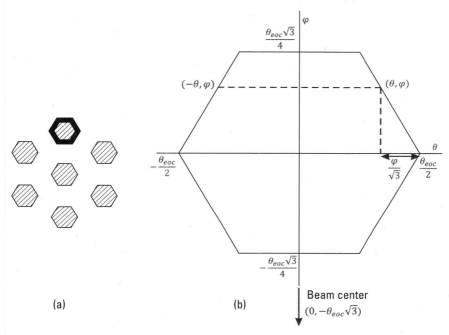

Figure 6.5 (a) The selected interference cell for analysis in the first order interference cluster and (b) the associated geometry assuming straight line contours for the antenna pattern within the cell.

Thus, the PDF describing the spatial probability distribution of being on an isolevel line within the cell can be put as

$$f_\Phi(\varphi) = \frac{8\theta_{eoc}\sqrt{3} - 16|\varphi|}{9\theta_{eoc}^2}$$

$$\text{for } |\varphi| \le \frac{\theta_{eoc}\sqrt{3}}{4}$$

$$f_\Phi(\varphi) = 0$$

$$\text{for } |\varphi| > \frac{\theta_{eoc}\sqrt{3}}{4}$$

(6.8)

This is plotted in Figure 6.6 together with the appropriate antenna pattern in the φ-axis. With the hypothesis of the analysis, the PDF is only a function of φ.

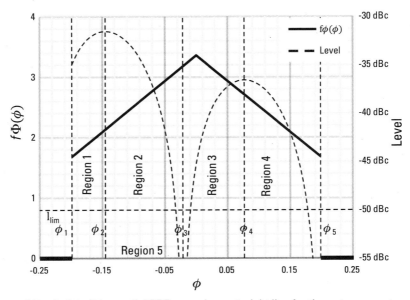

Figure 6.6 A plot of the spatial PDF assuming a straight line for the antenna contour over a cell and the interference pattern in the φ-axis on the decibel scale and the partitioning of the pattern to be used for the transformation.

In general [5, 6], the PDF for a continuous random variable X has three characteristics, described as follows:

1. The total area of the PDF is unity, or

$$\int_{-\infty}^{\infty} f_X(x)\,dx = 1 \tag{6.9}$$

2. The function is single-valued and nonnegative for all values of x.
3. For all defined values of a and b where $a \le b$

$$P[a \le X \le b] = \int_a^b f_X(x)\,dx \tag{6.10}$$

which implies that $P[X = a = b] = 0$.

Looking at Figure 6.6, we see that the PDF is a nonnegative even function.

The PDF in (6.8) can now be integrated to test for condition 2, and as can be seen in the following equation, the test is positive.

$$\int_{-\infty}^{\infty} f_\Phi(\varphi)\,d\varphi = 2\int_0^{\frac{\theta_{eoc}\sqrt{3}}{4}} \frac{8\theta_{eoc}\sqrt{3} - 16|\varphi|}{9\theta_{eoc}^2}\,d\varphi$$

$$= \left[16\frac{\theta_{eoc}\varphi\sqrt{3} - \varphi^2}{9\theta_{eoc}^2} \right]_0^{\frac{\theta_{eoc}\sqrt{3}}{4}} = 1$$

(6.11)

The spatial PDF is the basis, but the aim is to derive the distribution of the signal level that could be uplinked as interference from the cell. The spatial distribution describes the probability distribution of the points or locations within a given cell of being on a given antenna contour. The level PDF describes the probability distribution of the antenna performance (gain or directivity) associated with the ensemble of locations within the cell. Assuming a fixed EIRP for all terminals, the level PDF gives the power distribution of the interference contribution from that cell at the output port of the satellite uplink antenna through a link budget analysis. Thus, we need to transform the spatial distribution into the level distribution. We already know that the interval for level domain, the variable L, has an interval given by $(0, l_{SL2})$ where l_{SL2} is the peak level of the second sidelobe (i.e., $10^{31.5/10}$). We also note that the probability for corresponding segments of the spatial and level PDFs must be equal; employing (6.10), we can formulate this as

$$\int_{\varphi_A}^{\varphi_B} f_\Phi(\varphi)\,d\varphi = \int_{l_A}^{l_B} f_L(l)\,d\varphi$$

(6.12)

where $l = f(\varphi)$ and for $\varphi_B \geq \varphi_A$.

Note that there is a one-to-one mapping between φ_A and l_A, and φ_B and l_B. This implies that $l = f(\varphi)$ must be a monotonic function. Assuming that the function $l = f(\varphi)$ can be we can apply the method of transformations and as φ_A tends to φ_B (6.12) can be rewritten as [5]

$$f_X(l_x) = \frac{f_\Phi(\varphi_x)}{\left| \dfrac{dl_x}{d\varphi} \right|}$$

(6.13)

Note the modulus for the differential since a monotonous function can be increasing or decreasing but the PDF cannot be negative. Our antenna pattern within the cell is differentiable, not monotonic. However, it can be partitioned into monotonic segments.

Looking at Figure 6.6, we note that the antenna pattern within the cell can be decomposed into the following four monotonic regions:

- Region 1 from the lower cell edge to the second sidelobe peak, delimited by φ_1 and φ_2;
- Region 2 from the second sidelobe peak to the third null, delimited by φ_2 and φ_3;
- Region 3 from the third null to the third sidelobe peak, delimited by φ_3 and φ_4;
- Region 4 from the third sidelobe peak to the upper cell edge, delimited by φ_4 and φ_5.

Additionally, we define region 5 to cover all the areas of the above regions below an arbitrary low level of the pattern level, which we have taken to be −50 dB. Region 5 represents an area of the PDF where the interference level is too low to be significant and eases computational since it is deducted directly from the definitions of regions 1, 2, 3, and 4 in the φ domain. Figure 6.6 depicts the five regions.

The transformation as given in (6.13) is applied to the spatial PDF given in (6.8). This is performed employing computational techniques, and the result is shown in Figure 6.7. The data is given on the basis of a linear and a decibel scale. The basis for the mathematical analysis is linear units (not in decibels) but the number in decibels is more meaningful to engineers. Additionally, this is the PDF corresponding to a single-entry power distribution, and ultimately, we are interested in the sum of the interference power from the six cells at the uplink antenna output port.

The low-level simplification can be observed as the flat top at the low end of the distribution. This is very visible on the decibel version of the plot of Figure 6.8 (right), which should extend to negative infinity. The plot is truncated at −55 dB but as can be seen is flat beyond -50 dB. In real terms, this accounts for only about 10% of the total area as can be seen in Figure 6.8 (left), which actually goes to zero and at levels that need to be accounted for but where detail is not important.

The spikes may appear as odd but if we replot Figure 6.6 with the antenna pattern on a linear scale as shown in Figure 6.8, the explanation may become more evident. Although this may not be so familiar to engineers, it is the basis of the mathematical analysis. It will be seen that $d_{lx}/d\varphi$ of the antenna pattern is zero at the peaks of the second ($10^{-31.5/10}$) and third sidelobes ($10^{-36.6/10}$) and the third and fourth nulls toward zero. Thus, employing the transformation given in (6.13), the value of the PDF tends to infinity as we approach a peak or a null. However, (6.10) shows that the probability of being at these values

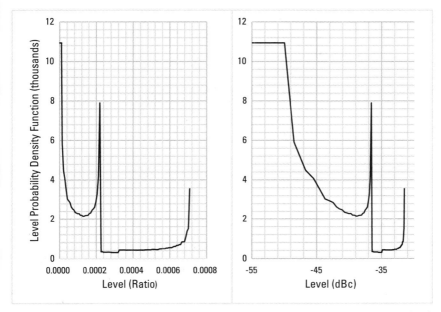

Figure 6.7 The level PDF over a hexagonal cell computed from the spatial PDF in (6.8) and shown in Figure 6.6. On the right the *x*-axis is in linear units, while on the left it is in decibels.

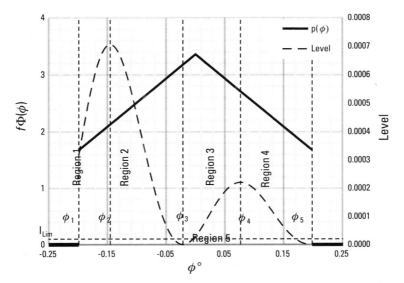

Figure 6.8 A plot of the PDF for the spatial distribution over a cell and the interference pattern in the φ-axis on a linear scale. This contains the same information as Figure 6.6; only the scale is different.

is zero, and we know from (6.12) that the integral including these singularities is finite since we know the value of the integral from the φ domain.

We can also look at the cumulative distribution function (CDF) of this single-entry interference signal; this is given in Figure 6.9. The equation gives us the probability that $L \leq l$ and is thus defined as

$$F_L(l) = \int_0^l f_L(l)\,d\varphi \tag{6.14}$$

This is particularly important in our interference scenario where we are interested in the probability that the interference level does not exceed a certain interference threshold and therefore a certain performance can be guaranteed. This can be used as a criterion for the link budget analysis.

We compare two plots in Figure 6.9, namely the plots entitled *straight-line model* and *data*. The other plots will be discussed later in this section after the specific models are presented. The CDF for the straight-line model corresponds to that produced in (6.14), while the data plot is the cumulative relative frequency of the data from Figure 6.4. The CDF for the straight-line model exhibits the same shape as the data although with a larger curvature below the

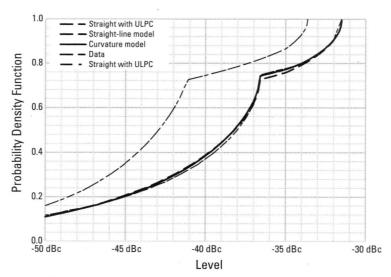

Figure 6.9 The CDF for a single-entry interference for the straight-line model, the curvature model and the straight-line model. Also shown is the cumulative relative frequency for the data produced in Figure 6.4.

third sidelobe level, which occurs at a slightly lower probability. This due to the straight-line simplification of spatial PDFs where the distribution is biased toward the high side and therefore deforms the curve toward the low side. At the third sidelobe level the value of the relative cumulative frequency of the data is about 75% while the value for the straight-line model is about 72%.

We can go a step further and assess the situation if the curvature of the isolevel contours of the antenna pattern is considered. The geometry is shown in Figure 6.10, which, to avoid confusion, maintains the same coordinate system as in the straight-line approach and the same reference cell [as in Figure 6.5 a)]. The following four regions can be identified in Figure 6.10:

- Region 1 ranges from when the contour just touches the cell side closest to the beam center until it is completely within the cell and just touches the side of the hexagon that bounds the contour. The

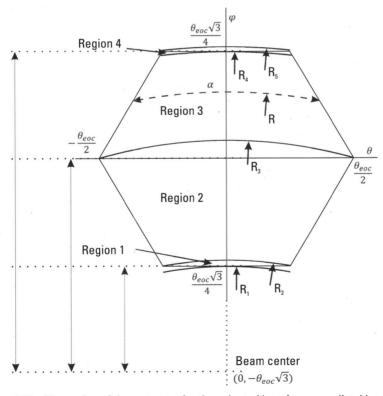

Figure 6.10 Illustration of the geometry for the selected interference cell, taking into account the curvature of the contours for the antenna pattern within the cell.

region lies between the limits given by arc produced by radii R_1 and R_2 within the cell.

- Region 2 is bound the lower right and left side of the hexagon and by R_2 and R_3.
- Region 3 is bound the upper right and left side of the hexagon and by R_3 and R_4.
- Region 4 is bound by the upper side of the hexagon and R_5. Because of the geometry considered this is very small.

R_1, R_2, R_3, R_4, and R_5 originate from the beam center.

Working through the geometry as illustrated in Figure 6.10, we get

$$R_1 = \frac{3\theta_{eoc}\sqrt{3}}{4}$$

$$R_2 = \frac{\theta_{eoc}\sqrt{7}}{2}$$

$$R_3 = \frac{\theta_{eoc}\sqrt{13}}{2} \tag{6.15}$$

$$R_4 = \frac{5\theta_{eoc}\sqrt{3}}{4}$$

$$R_5 = \frac{\theta_{eoc}\sqrt{19}}{2}$$

In general, the relationship between the coordinate within the cell and the beam radius is given by

$$\left(\varphi + \theta_{eoc}\sqrt{3}\right)^2 + \left(\frac{\theta_{eoc}}{2} - \frac{|\varphi|}{\sqrt{3}}\right)^2 - R^2 = 0 \tag{6.16}$$

$$\text{for } R_4 \geq R \geq R_3$$

And using the quadratic solution

$$\varphi = \frac{-5\theta_{eoc}\sqrt{3} + \sqrt{-81\theta_{eoc} + 48R^2}}{8}$$

$$\text{for } R_2 \leq R \leq R_3$$

$$\varphi = \frac{-7\theta_{eoc}\sqrt{3} + \sqrt{-9\theta_{eoc} + 48R^2}}{8} \tag{6.17}$$

$$\text{for } R_3 \leq R \leq R_5$$

We retain the positive addition of the quadratic solution since it represents the intersection of the circle with radius R with the part of the line covering the sides of the hexagon.

The isolevel contour is now the arc within the cell, the part of the circle within the cell. The length of the arc can now be expressed as

$$\alpha = 2R\cos^{-1}\frac{3\theta_{eoc}\sqrt{3}}{4R}$$

for $R_1 \leq R \leq R_2$, i.e., Region 1

$$\alpha = 2R\cos^{-1}\frac{\varphi + \theta_{eoc}\sqrt{3}}{R}$$

for $R_2 \leq R \leq R_4$, i.e., for Regions 2 and 3 $\qquad$ (6.18)

$$\alpha = 2R\left\{\cos^{-1}\frac{\varphi + \theta_{eoc}}{R} + \cos^{-1}\frac{5\theta_{eoc}\sqrt{3}}{4R}\right\}$$

for $R_4 \leq R \leq R_5$, i.e., for Region 4

This spatial PDF that describes the probability distribution of an ensemble of points or locations within the cell is on an isolevel arc within the cell and can now be expressed as the ratio of the linear length of the arc within the cell to the cell area. Therefore, using (6.18) and (5.12), we can deduce the spatial PDF as function of a continuous variable A which can be put as

$$f_A(\alpha) = \frac{16R}{3\theta_{eoc}\sqrt{3}}\cos^{-1}\frac{3\theta_{eoc}\sqrt{3}}{4R}$$

for $R_1 \leq R \leq R_2$, i.e., Region 1

$$f_A(\alpha) = \frac{16R}{3\theta_{eoc}\sqrt{3}}\cos^{-1}\frac{\varphi + \theta_{eoc}\sqrt{3}}{R}$$

for $R_2 \leq R \leq R_4$, i.e., for Regions 2 and 3 $\qquad$ (6.19)

$$f_A(\alpha) = \frac{16R}{3\theta_{eoc}\sqrt{3}}\left\{\cos^{-1}\frac{\varphi + \theta_{eoc}}{R} + \cos^{-1}\frac{5\theta_{eoc}\sqrt{3}}{4R}\right\}$$

for $R_4 \leq R \leq R_5$, i.e., for Region 4

The three conditions characterizing a PDF are given (6.9) and (6.10) together with the requirement that the function is single-valued and non-negative for all values of x. We confirm that the integral of the PDF over the defined interval using the appropriated definitions for $f_A(\alpha)$ for regions 1–4 as given in (6.19) is

$$\int_{R_1}^{R_2} f_A(\alpha)\,dR + \int_{R_2}^{R_3} f_A(\alpha)\,dR + \int_{R_3}^{R_4} f_A(\alpha)\,dR + \int_{R_4}^{R_5} f_A(\alpha)\,dR = 1 \quad (6.20)$$

Figure 6.11 depicts (6.19) together with the antenna pattern. It will be noticed that the function is single-valued and nonnegative for all values. It shows that the spatial PDF is no longer an even function about the center of the cell but is skewed toward the positive side of the φ-axis. This is because, as shown in Figure 6.10, the curvature introduces a bias in this direction. In the straight-line model, the equivalents of regions 1 and 2 were equal and symmetric. In the curvature model, region 2 is larger than region 3 as shown in Figure 6.10 and thus the skew. At the negative edge of the cell, we can also observe an addition segment. This is due to region 1. Region 4 is too small to have a visual effect.

At this point, we can develop the level PDF in the same way we did for the straight-line approach. This implies the same segmentation approach principally based on four monotonic intervals of the antenna pattern and the low-level region. The same transformation is given in (5.13). In fact, the same computational process is used, and the only difference is the definition

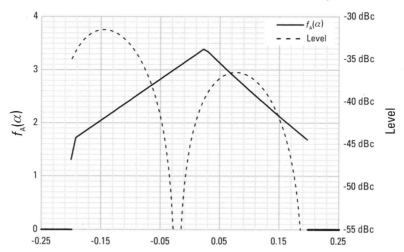

Figure 6.11 A plot of the PDF for the spatial distribution, taking into account the curvature of the antenna contour within the cell and the interference pattern in the φ-axis on the decibel scale.

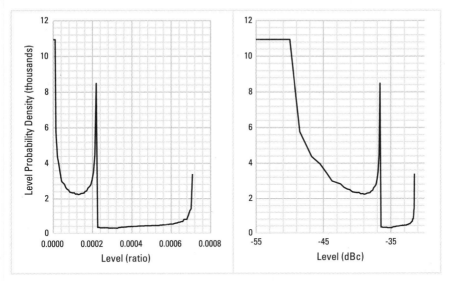

Figure 6.12   The level PDF over a hexagonal cell computed from the spatial PDF in (6.19) and shown in Figure 6.11. On the right the *x*-axis is in linear units, while on the left it is in decibels.

of the spatial PDF. Figure 6.12 shows the resulting level PDF; the plots have the same structure as those in the case with the straight-line assumption, and the same comments apply.

From the level PDF we can now produce the level CDF as shown in Figure 6.9. The correlation is very good for the curvature model and the data over the full interval with the point at the third sidelobe peak almost coincidental at 75%. This is an improvement with respect to the straight-line model, which remains a simple model that can be easier to use and good enough for practical purposes.

Now that we manage to characterize the single-entry interference model, we need to extend it to the six-cell interference cluster. The PDF for two independent variables is given by the convolution of their respective PDFs. By extension since we are dealing with six cells, the resulting level PDF is the convolution of the six PDFs. We assume that the interference contributions from the six cells are independent, and in our simplified model, the six PDFs are identical. The convolution of two PDFs is defined as [5]

$$f_X(z) \otimes f_Y(z) = \int_{-\infty}^{\infty} f_X(w) f_Y(z-w)\, dw \qquad (6.21)$$

where $\otimes$ represents the convolution operator.

It should be noted that the convolution operation is commutative so that the order of the operands is irrelevant.

By extension, for the six-cell interference system, we can express the aggregate six-entry interference level as the convolution of the six probability density functions from the six cells.

$$f_{L_T}(l) = f_{L_1}(l) \otimes f_{L_2}(l) \otimes f_{L_3}(l) \otimes f_{L_4}(l) \otimes f_{L_5}(l) \otimes f_{L_6}(l)$$

$$\left(\int_{-\infty}^{\infty} \left\{ \int_{-\infty}^{\infty} \left[\int_{-\infty}^{\infty} \left(\int_{-\infty}^{\infty} f_{L_1}(x_1) f_{L_2}(x_2 - x_1) dx_1 \right) f_{L_3}(x_3 - x_2) dx_2 \right] \right. \right. \quad (6.22)$$

$$\left. \left. f_{L_4}(x_4 - x_3) dx_3 \right\} f_{L_5}(x_5 - x_4) dx_4 \right) f_{L_6}(l - x_5) dx_5$$

where x_1, x_2, x_3, x_4, and x_5 are dummy variables, since they disappear after the operation.

Equation (6.22) may look daunting, but dealing with it numerically is manageable. This is illustrated in Figure 6.13. Notice that the single-entry distribution spans from zero to $10^{-31.5/10}$. This is logical since this is the dynamic range of the single-entry contribution. With multiple entries, the lower bound is fixed to zero. The antenna gain or directivity as a numerical ratio cannot be negative. For a dual entry, the upper bound is twice that of the single-entry since there is a finite probability that there is contribution in the range close to twice that of a single-entry. Thus, as the order of the convolution increases, the PDF gets smoother, and at the sixth convolution, the upper bound stretches to six times that of the single-entry. The extension with each convolution process is also demonstrated in Figure 6.13.

Note that the convolution process has a smoothing effect and that the spikes of the single-entry PDF disappear after the first convolution with the fifth being smooth.

The final step of the interference analysis is to generate the level CDF of the six-cell scenario, shown in Figure 6.14. This can be used to establish a performance criterion (i.e., the percentage probability that the six-cell interference level is below a certain threshold).

We can now proceed further and introduce profiling of the EIRP across the cell so that the uplink EIRP from the terminal is a function of where it is within the cell to compensate for the G/T variation. The main objective would be that the level of the various carriers originating across the cell that would appear at an equal level at the TWTA input. This ensures that the C/I due to the nonlinear impairment of the TWTA is not degraded for the small carriers.

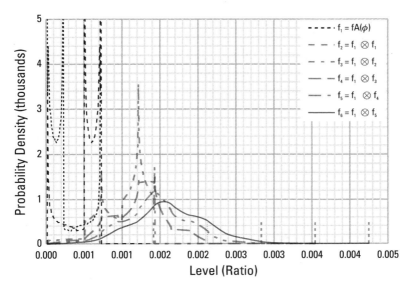

Figure 6.13 Plot of the level PDF and the convolutions up to six times leading to the total level PDF, giving the probability distribution of the aggregate six-entry interference.

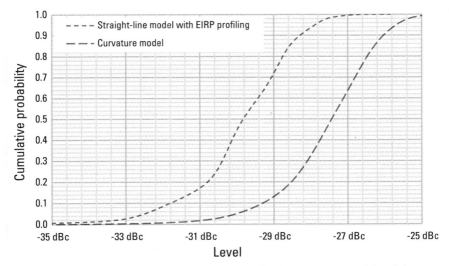

Figure 6.14 The level CDF of the six-cell scenario for the curvature model and the straight-line model with EIRP profiling across the cell.

In fact, what is important is that the EIRP density at the TWTA input is constant. Thus, a possible implementation is to maintain a nominally constant EIRP and change the bandwidth so that a terminal at the center of the beam employs a higher symbol rate than a user at the edge of the beam.

For the analysis of this scenario, we use a simple approach employing the straight-line model and estimating an average compensation level on a line basis. Figure 6.15 is based on Figure 6.6. The spatial distribution and the antenna level are identical. However, a compensation is introduced so that the uplink EIRP is a function of the terminal location within the cell. This is used to produce a modified antenna pattern as shown in Figure 6.15.

The same process is pursued as indicated above, and Figure 6.14 shows the CDF. The EIRP profiling results in a shift in the S-curve to the left. In fact, the simple view is that the shift at the high end is equal to the difference between the second sidelobe peak and the modified version (i.e., about 2 dB).

This section provides an overview of the topological mechanism of the generation of copolar cell interference. Unlike the forward link, this assumes a probabilistic nature because of the randomness of the sites of the sources.

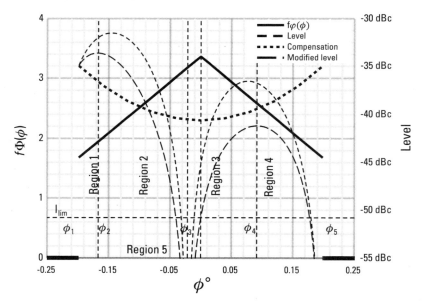

Figure 6.15　Illustration of the EIRP profile model using the straight-line approach and showing the antenna pattern, the compensation curve, and the equivalent antenna pattern. Note that the compensation is on the decibel scale but only in a relative fashion. The absolute decibel scale does not apply.

The analysis is performed for one interfering cell, and the results are depicted. The analysis is then extended to a first-order cluster of six copolar cells. Two approaches are presented: one where the contours of the spacecraft uplink antenna within the interference cell is approximated by a straight line and another where the curvature of the contour is considered. Two scenarios are given: one where the terminal EIRP is flat and independent of its location within the cell and another where the terminal EIRP is compensated across the cell so that the power into the satellite TWTA is independent of the terminal location. CDFs are given for the interference level into the LNA.

6.5 Access Scheme

Recall that the signal structure of the return is different from that of the forward; this section considers this difference and discusses the access scheme as defined in DVB-RCS2, an internationally recognized standard. This section also addresses the ACM scheme for the return link.

Most return link systems are based on MF-TDMA, because it allows for the dynamic sharing of bandwidth resources. In the return link, the terminal originates the uplink traffic, which tends to be bursty. For a cost-effective terminal, the EIRP is a critical element. There is a marketing drive towards small aperture terminals but the SSPA is a cost item. Thus the economic push is for the lowest possible power for the given application. MF-TDMA offers a good compromise for satellite return links, because it provides the most bandwidth and the greatest overall efficiency and service quality, while also allowing the dynamic sharing of that bandwidth among a large population of terminals.

Most systems tend to be proprietary but have a strong resemblance to the best-known open system, digital video broadcasting-return channel via satellite (DVB-RCS), at least at the lower layers. The standard was first published by ETSI in 2000 as EN 301 790 V1.2.2, and there have been several issues, with the last in 2009. In the 2008 version, DVB extended the specifications to cover mobile terminals as version 1.5.1 [7]. This allowed the standard to adapt to different market segments from small to large networks and from fixed to mobile terminals.

The first-generation DVB-RCS [7] was limited to QPSK modulation and supported two coding schemes: concatenated and turbo coding.

Concatenated coding includes an outer code that is based on the Reed-Solomon RS (255, 239, 8) and an inner code that is a punctured convolutional code with $K = 7$ based on a rate 1/2 mother code so that 1/2, 2/3, 3/4, 5/6, and 7/8 are supported. The inner and the outer codes are bypassable.

The turbo encoder uses a double binary circular recursive systematic convolutional code to create rates 1/3, 2/3, 2/5, 1/2, 3/4, 4/5, and 6/7 through puncturing.

In both cases, error detection can be aided through cyclic redundancy check (CRC).

The specification sets the roll-off factor to 35% and the terminal EIRP step at 0.5 dB but leaves the EIRP range up to the manufacturer.

The first version of the second generation of the DVB-RCS2 was published in 2012, and the last version is in draft form at the time of writing. The specification of the second generation has been expanded to a multipart specification [8–12]. The main objective of the second generation is to open the door for interoperability between terminals and hubs, thus including network, management and control functions.

The MF-TDMA of DVB-RCS2 is structured around the super-frame definition, which is organized in two dimensions: frequency and time. This is illustrated in Figure 6.16 where the horizontal axis is time, and the vertical axis is frequency. A number of center frequencies are designated so that a time frame structure can exist on each center frequency. The basic element is the bandwidth time unit (BTU). It represents the shortest possible burst at one of the possible frequencies. The elemental carrier bandwidth is undefined and left to the system implementation but is compatible with carrier throughput in excess of 10 Mbps in fixed Ka-broadband applications.

A frame could be composed of time slots that are a single BTU long labeled as A in Figure 6.16. Time slots could also have durations that are an integer multiple duration of BTUs designated by the labels B and C for twice or six times the BTU duration as shown in Figure 6.16. The duration of a super-frame is specified as ranging in duration from 25 to 750 ms as a minimum. Transitions between time slots occur at predefined possible epochs. The number of time slots in a super-frame is limited to 2,047. The first level of flexibility to increase capacity is to increase the time slot duration as it does not necessitate extra RF power from the terminal, but it is possible to increase the capacity by occupying more than one carrier. No changes occur during an active super-frame.

In DVB-RCS2, besides QPSK, other modulation schemes are included, namely continuous phase modulation (CPM), 8PSK, and 16QAM where amplifiers can be run closer to saturation. CPM is claimed to be more efficient than linear modulation for a given SSPA [13]. This allows for cost optimization for the RF resource in the terminal.

The FEC for QPSK, 8PSK, and 16QAM is a 16-state turbo code, commonly called turbo-phi. The coding gain of the turbo-phi scheme varies

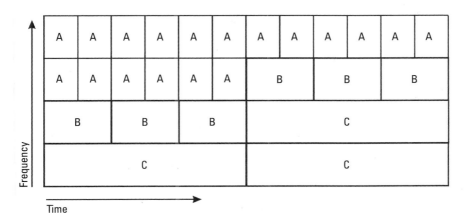

Figure 6.16 Illustration of the super-frame implementation of the MF-TDMA structure of DVB-RCS2.

according to the modulation, coding rate, and burst sizes with respect to that of the RCS1 turbo code, but it approaches values around 1 dB at low-BER values. This extra coding gain can be used at system level. The rates employed are 1/3, 1/2, 2/3, 3/4, and 5/6 as part of the reference waveforms for specification for linear modulation bursts. Table 6.1 details the performance for linear modulation with burst sizes of 266, 536, 1,616, and 3,236 symbols with the appropriate modulation and coding.

Convolutional coding with $K = 3$ or 4 is specified in conjunction with CPM with rate 1/2, 2/3, 3/4, 4/5, and 6/7. Table 6.2 details the performance for CPM bursts containing 400, 1,024, and 1,504 information bits with the appropriate modulation and coding.

Combinations of modulation and coding are used in a timeslot, allowing per-timeslot ACM for more granular and more flexible link adaptation. ACM on the return link allows the best use of the RF amplifier with the best symbol rate and QoS.

The following three modes of EIRP are available:

- Constant EIRP adjusted by network control center (NCC);
- Autonomous EIRP control based on E_s/N_0 reported by NCC;
- Constant power spectrum density adjusted by NCC.

The roll-off factor is reduced to 20%.

DVB-RCS2 offers considerable advantages with respect to DVB-RCS since offers extended possibilities in the ACM operation and also other network function.

Table 6.1
The Performance of DVB-RCS2 Established from Table A-1 of [9] and Tables 10.4–10.8 of [11] for Linear Modulation Bursts

Waveform ID	Burst Size (Sym.)	Guard (Sym.)	Payload (bits)	Mod.	FEC	Eff. (Bits/ Sym)	E_s/N_0 @ PER $= 10^{-5}$	E_s/N_0 @ PER $= 10^{-3}$
44	266	4	408	QPSK	5/6	1.51	7.30	6.52
45	266	4	440	8-PSK	2/3	1.63	8.71	8.20
46	266	4	496	8-PSK	3/4	1.84	10.04	9.41
47	266	4	552	8-PSK	5/6	2.04	11.59	10.83
48	266	4	672	16-QAM	3/4	2.49	11.73	11.24
49	266	4	744	16-QAM	5/6	2.76	13.18	12.56
3	536	4	304	QPSK	1/3	0.56	0.22	-0.27
4	536	4	472	QPSK	1/2	0.87	2.34	1.92
5	536	4	680	QPSK	2/3	1.26	4.29	3.90
6	536	4	768	QPSK	3/4	1.42	5.36	4.93
7	536	4	864	QPSK	5/6	1.6	6.68	6.11
8	536	4	920	8PSK	2/3	1.7	8.08	7.71
9	536	4	1040	8PSK	3/4	1.93	9.31	8.90
10	536	4	1152	8PSK	5/6	2.13	10.85	10.43
11	536	4	1400	16QAM	3/4	2.59	11.17	10.83
12	536	4	1552	16QAM	5/6	2.87	12.56	12.16
13	1616	4	984	QPSK	1/3	0.61	-0.51	-0.80
14	1616	4	1504	QPSK	1/2	0.93	1.71	1.49
15	1616	4	2112	QPSK	2/3	1.3	3.69	3.46
16	1616	4	2384	QPSK	3/4	1.47	4.73	4.50
17	1616	4	2664	QPSK	5/6	1.64	5.94	5.64
18	1616	4	2840	8PSK	2/3	1.75	7.49	7.29
19	1616	4	3200	8PSK	3/4	1.98	8.77	8.56
20	1616	4	3552	8PSK	5/6	2.19	10.23	10.02
21	1616	4	4312	16QAM	3/4	2.66	10.72	10.55
22	1616	4	4792	16QAM	5/6	2.96	12.04	11.86
42	3236	4	984	BPSK	1/3	0.3	−3.52	−3.81
43	3236	4	1504	BPSK	1/2	0.46	−1.30	−1.53

Table 6.2
The Performance of DVB-RCS2 Compiled from Table A-2 of [9] and Tables 10.10–10.12 of [11] for CPM Waveforms

Waveform ID	Info. bits	Normalized Carrier Spacing (fT)	Pulse Shape (α_{rc})	Mod. index (h)	Code rate	Spectral Efficiency b/s/Hz	E_s/N_0 @ PER $= 10^{-5}$	E_s/N_0 @ PER $= 10^{-3}$
3	400	2.000	0.980	2/5	1/2	0.50	2.2	2.8
4	400	1.333	0.750	1/3	1/2	0.75	2.8	3.3
5	400	1.210	0.750	2/7	2/3	1.10	3.7	4.4
6	400	1.067	0.750	2/7	2/3	1.25	4.4	5.2
7	400	1.067	0.750	1/4	4/5	1.50	6.1	7.2
8	400	0.974	0.625	1/5	6/7	1.80	9.2	11.1
9	1024	2.000	0.980	2/5	1/2	0.50	1.8	2.0
10	1024	1.333	0.750	1/3	1/2	0.75	2.3	2.6
11	1024	1.210	0.750	2/7	2/3	1.10	3.2	3.5
12	1024	1.067	0.750	2/7	2/3	1.25	3.6	4.2
13	1024	1.067	0.750	1/4	4/5	1.50	5.4	6.0
15	1504	2.000	0.980	2/5	1/2	0.50	1.6	1.8
16	1504	1.333	0.750	1/3	1/2	0.75	2.2	2.4
17	1504	1.210	0.750	2/7	2/3	1.10	3.0	3.4
18	1504	1.067	0.750	2/7	2/3	1.25	3.5	3.9
19	1504	1.067	0.750	1/4	4/5	1.50	5.2	5.6
20	1504	0.974	0.625	1/5	6/7	1.80	8.1	9.0

6.6 The Return Link Budget

This section develops a link budget assessment that ultimately provides the return capacity. As with the forward link, for the return link, we assume that the link budget performance is asymmetric and, in this case, dominated by the uplink performance. The downlink is to the gateway, which means that the downlink can be dimensioned such that the performance is principally determined by the uplink.

Table 6.3 details the reference link budget at the cell center and the underlying assumptions for the ground segment. The antenna apertures are, of course, as defined for the forward link, since typically the same antenna aperture is used for both the uplink and downlink for the terminal and the gateway, respectively. The maximum RF power of terminal SSPA is assumed to be 2W. The payload performance is as discussed earlier, namely the satellite G/T at the boresight is 26 dB/K. The downlink is assumed to be dimensioned similarly to the forward downlink. This need not be the case. Considering the size of the gateway, the ensuing G/T is very good.

The downlink performance is constant under static propagation conditions. The uplink is a function of the location of the terminal in the cell. Looking at the link budget, as expected, the link is very asymmetric since the downlink C/N_0 is considerably better than the uplink. The link budget performance is thus determined by the uplink.

The reference link provides the fixed elements for the estimation of the performance over the cell. The $\left(\dfrac{C}{N+I}\right)_{RTN}$ that the gateway receives can thus be expressed as

$$\left(\frac{C}{N+I}\right)_{RTN} = -10\log_{10}\left(\begin{array}{l} 10^{\frac{-(C/N)_{RTN,UL}}{10}} + 10^{\frac{-(C/I_C)_{RTN,UL}}{10}} + 10^{\frac{-(C/I_X)_{RTN,UL}}{10}} \\ + 10^{\frac{-(C/N)_{Rptr}}{10}} + 10^{\frac{-(C/I)_{RTN,Rptr}}{10}} + 10^{\frac{-(C/N)_{RTN,DL}}{10}} \end{array}\right)$$

$$(6.23)$$

where:

$(C/N)_{RTN,UL}$ is the uplink carrier-to-noise ratio in decibels from a given terminal, which is a function of the radial distance from the beam boresight or cell center;

$(C/I_C)_{RTN,UL}$ is the uplink cofrequency copolar C/I in decibels where the interference level is a function of the CDF of the six-cell cluster as discussed in the above section, and the carrier level from a given terminal is a function of the radial distance from the beam boresight or cell center;

$(C/I_X)_{RTN,UL}$ is the uplink cross-polar C/I in decibels of the cofrequency cross-polar principally determined by the two adjacent cells;

$(C/N)_{Rptr}$ is the repeater impairment in decibels, which is principally the NPR that is applicable for a given TWTA used for the channel;

$(C/I)_{RTN,Rptr}$ is the repeater impairment due to the forward link;

Table 6.3
Reference Clear-Sky Link Budget for the Return Link at the Beam Boresight

Uplink			
Earth Station		**Uplink**	
Frequency	29.50 GHz	Uplink EIRP	46.9 dBW
HPA	2.0W	Atmospheric loss	2.00 dB
Output losses	0.5 dB	Range	37236 km
OBO	0.5 dB	Free-space loss	213.3 dB
Antenna diameter	0.70m	Spreading loss	162.4 dB
Antenna beamwidth	1.02°	Satellite G/T	26.1 dB/K
Antenna gain	44.9 dBi	Input flux density	−117.5 dBW/m²
Uplink EIRP	46.9 dBW	C/N_0	86.4 dBHz
Downlink			
Earth Station		**Downlink**	
Frequency	19.00 GHz	Satellite EIRP	66.8 dBW
Antenna diameter	9.10m	OBO	5.0 dB
Antenna beamwidth	0.12°	Total bandwidth	750.0 MHz
Antenna gain	63.4 dBi	Carrier spacing	6.0 MHz
Antenna temperature	40.0K	Number of carriers possible	125
Receiver noise factor	3.0 dB	Pointing loss	1.5 dB
Receiver noise temperature	288.6K	Atmospheric loss	1.80 dB
System temperature	328.6K	Range	37236 km
	25.2 dBK	Free-space loss	209.5 dB
G/T	38.2 dB/K	Earth station G/T	38.2 dB/K
		C/N_0	94.9 dB
Total			
		C/N_0	85.8 dB
		C/N	18.0 dB

$(C/N)_{\text{RTN,DL}}$ is the downlink carrier-to-noise ratio in decibels to a given gateway.

The $(C/I_C)_{\text{RTN,UL}}$ is estimated from Figure 6.14. EIRP profiling is assumed with the objective that the uplink $\left(\dfrac{C}{N+I} \right)$ is constant whether the signal originates from a terminal situated at the cell center or at the edge. If we adopt the criterion that the interference level should not be exceeded for 99.5% of the cases, we arrive at a figure of 27.1 dB down with respect to the boresight level.

$(C/I_X)_{\text{RTN,UL}}$ is assumed to be 25 dB per entry. Since in our simple model a dual entry is assumed for this contribution, 22 dB is used.

It should be noted that for the return link, the downlink of a MF-TDMA signal is a multiplex of several carriers with traffic from a number of cells over a large bandwidth. Thus, the EIRP profiling over the cell is assumed to optimize the TWTA's nonlinear performance. This means that the $(C/N)_{\text{Rptr}}$ is given by the NPR of the TWT operated at the appropriate output back-off (OBO). In our case, we assume a 5-dB OBO to provide an NPR of 20.5 dB where no ULPC is employed.

The $(C/I)_{\text{RTN,Rptr}}$ considers the interference that may be generated within the repeater from the forward payload, such as noise, because of the proximity of the spectrum used; it is considered to be negligible in this analysis.

The analysis is performed for downlink channel bandwidths of 500 MHz; 750 MHz; 1,000 MHz; 1,250 MHz; and 1,500 MHz and the results are shown in Figure 6.17. For simplicity, it is assumed that linear modulation is employed and that all bursts are 1,616-symbols-long. Looking at Figure 6.17, it will be noticed that the capacity is represented by a straight line. This is because the uplink is composed of a number of signals, and as more bandwidth is available, more uplink carriers can be accommodated, but the uplink budget is constant on power density basis. In our case, the uplink performance is the limiting link so that the degradation due to the downlink is small. This is demonstrated by the spectral efficiency in Figure 6.17, which only has a slight negative gradient. Also looking at the modulation split, there is a slight shift toward the lower-order modulation.

It will also be noticed that the capacity is almost linear with the available downlink bandwidth. This is because in a MF-TDMA uplink the EIRP density is independent of bandwidth. As more bandwidth becomes available, more terminals can be accommodated resulting in constant EIRP density. The downlink is power-constrained, and if the satellite system is adequately dimensioned, it is not the limiting factor.

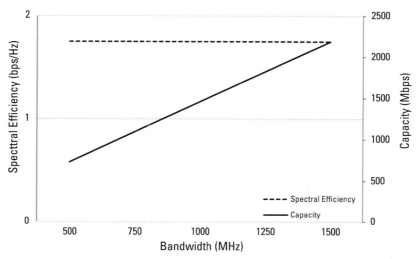

Figure 6.17 Cell performance showing the split between the modulation schemes and the capacity and spectral efficiency.

Since we have assumed that the terminal bandwidth is adjusted so that the power density at the TWTA input is equalized, then the downlink EIRP density is also equalized so that $\left(\dfrac{C}{N+I}\right)_{\text{RTN,DL}}$ for each carrier is also constant. Of course, as the system bandwidth increases, the EIRP density decreases. Whether this has an overall effect is a function of the difference between $(C/N+1)_{\text{RTN,UL}}$ and $(C/N+1)_{\text{RTN,DL}}$. As the bandwidth increases further, or as the difference between the uplink and the downlink decreases, the link budget function will deviate from the linear function until the overall link budget becomes limited by the downlink. Of course, systems are normally designed not to be limited by the downlink.

Having analyzed one cell, the analysis can be replicated to cover all the cells to give the system capacity.

Chapter 5 covers a technique to estimate the capacity of the forward link. This chapter extends that analysis to the return link. Together, Chapters 5 and 6 fully characterize the system in terms of the system capacity

References

[1]   Lo, Y. T., and S. W. Lee, *Antenna Handbook, Volume II, Antenna Theory*, New York: Van Nostrand Reinhold, 1993.

[2] Volakis, J., *Antenna Engineering Handbook* (Fourth Edition), McGraw-Hill Education, 2007.

[3] Cwik, T., and V. Jamnejad, "Beam Squint Due to Circular Polarization in a Beam-Waveguide Antenna," *TDA Progress Report 42-128*, 15 February 1997, https://pdfs .semanticscholar.org/fda7/e0bdfb6dea57d2497379a703cb13b55ac9c5.pdf.

[4] Fenech, H., et al., "G/T Predictions of Communication Satellites Based on a New Earth Brightness Model," *International Journal of Satellite Communications*, Volume 13, Issue 5, September/October 1995, pp. 367–376.

[5] Pishro-Nik, H., *Introduction to Probability, Statistics, and Random Processes*, available at https://www.probabilitycourse.com, Kappa Research LLC, 2014, https://www .probabilitycourse.com/.

[6] Vardeman, S. P., *Statistics for Engineering Problem Solving*, IEEE Press, PWS Publishing Company, 1994.

[7] *Digital Video Broadcasting (DVB); Interaction Channel for Satellite Distribution Systems*, ETSI EN 301 790 V1.5.1 (2009-05) European Standard (Telecommunications series), European Telecommunications Standards Institute (ETSI), Sophia Antipolis, France.

[8] *Digital Video Broadcasting (DVB); Second Generation DVB Interactive Satellite System (DVB-RCS2); Part 1: Overview and System Level Specification DVB*, DVB Bluebook A155-1, (Draft TS 101 545-1), January 2019, European Telecommunications Standards Institute (ETSI), Sophia Antipolis, France.

[9] *Digital Video Broadcasting (DVB); Second Generation DVB Interactive Satellite System (DVB-RCS2); Part 2: Lower Layers for Satellite Standard DVB*, DVB Bluebook A155-2 (Draft EN 301 545-2 V1.3.1), February 2019, European Telecommunications Standards Institute (ETSI), Sophia Antipolis, France.

[10] *Second Generation DVB Interactive Satellite System (DVB-RCS2) Part 3: Higher Layers Satellite Specification*, DVB BlueBook A155-3 (draft TS 101 545-3 V1.3.1), February 2020, European Telecommunications Standards Institute (ETSI), Sophia Antipolis, France.

[11] *Digital Video Broadcasting (DVB); Second Generation DVB Interactive Satellite System (DVB-RCS2); Part 4: Guidelines for Implementation and Use of EN 301 545-2*, ETSI TR 101 545-4 V1.1.1 (2014-04).

[12] *Digital Video Broadcasting (DVB); Second Generation DVB Interactive Satellite System (DVB-RCS2); Part 5: Guidelines for the Implementation and Use of TS 101 545-3*, ETSI TR 101 545-5 V1.1.1 (2014-04), European Telecommunications Standards Institute (ETSI), Sophia Antipolis, France.

[13] Skinnemoen, H., et al., "DVB-RCS2 Overview, Special Issue on DVB-RCS2," *International Journal on Satellite Communications and Networking*, Volume 31, Issue 5, September/October 2013, pp. 199–276.

7

Current GEO HTS Systems

7.1 Introduction

The genesis of the HTS is difficult to establish. One of the first satellites with the characteristics of an HTS satellite is the Telesat satellite, Anik F2, which was based on a Boeing 702 platform and launched in July 2004 on Ariane 5G+ at 111.1°W. The satellite is a multimission satellite including payloads in the C-band for satellite trunking into remote communities and the Ku-band over North America for enterprise and government mobility applications. The Ka-band payload operates carries 50 TWTAs at 90-W RF each and has 45 beams over North America for broadband services. The satellite was quoted to deliver 2 Gbps at the Ka-band [1].

The following year in August 2005, Thiacom launched IPStar 1 on an Ariane 5G with a launch mass of 6,505 kg to the orbital position 119.5°E. It was hailed as the heaviest satellite launched at the time. The satellite, also known as Thaicom 4, was manufactured by SSL (now Maxar). It has 84 Ku-band spot beams, three shaped beams, seven broadcast beams, and 18 Ka-band spot gateway (uplink) beams to provide voice, video, and broadband Internet services to 14 countries including India, Thailand, Japan, Indonesia, and Australia. The HTS system operates in the Ku-band for the users and the Ka-band for the gateways delivering 38.4 Gbps of capacity [2, 3].

This chapter discusses the operational geostationary satellites that are in orbit. It is not possible to cover all of the many systems here, but the discussion should give readers a flavor of some of the operational systems and their applications.

7.2   The European Telecommunications Satellite Organization (Eutelsat) HTS Systems

Eutelsat was originally set up in 1977 by 17 European countries as an intergovernmental organization (IGO). Its role was to develop and operate a satellite-based telecommunications infrastructure for Europe. The convention establishing the organization was opened for signature in July 1982 and entered into force on 1 September 1985.

In 1983, EUTELSAT launched its first satellite, Eutelsat 1 F1, to be used for telecommunications and TV distribution. EUTELSAT rapidly developed its infrastructure beyond Europe to cover the Middle East, the African continent, and large parts of Asia and the Americas.

EUTELSAT was the first satellite operator in Europe to broadcast television channels direct-to-home. It developed its premium neighborhood of the Hot Bird constellation, starting Hot Bird 1 in 1995 to offer capacity that would attract thousands of channels to the same orbital location with a large catchment area for consumer satellite TV. Broadcasting still contributed 62% of the revenue. The main sectors that use HTSs are fixed broadband and mobile connectivity, each accounting for 6% revenue [4].

Eutelsat's first step in the direction of broadband was with Hot Bird 6, which was launched on 21 August 2002 at 13° East. Hot Bird 6 was mainly a broadcasting satellite but included a Ka-band payload with four downlink spot beams and one contoured uplink beam. The coverage was Central Europe. It included the Skyplex regenerative system so that combinations bounded by 18 carriers at 2 Mbps each and six carriers at 6 Mbps could be uplinked from each beam and downlinked as a DVB-S stream. The system, which supported star and meshed networks, led to a service that was call SkyplexData.

A second step was e-Bird, which was launched in 27 September 2003 to be positioned at 31° East. Although e-Bird, which later became known as Eutelsat 31A, was the smallest satellite in Eutelsat's fleet with a launch mass of 1,530 kg, it was dedicated for multimedia and capable of supporting four return channels with a bandwidth of 108 MHz and 16 forward channels at 33 MHz. The satellite, which operated in the Ku-band and had four coverages, was finally used for broadcasting.

Neither of these two satellites garnered the critical commercial mass, and they probably were ahead of their time since the supporting ground segment was not sufficiently developed.

7.2.1   KA-SAT

The Eutelsat satellite KA-SAT can be considered as the first of its generation of Ka-band HTS satellites. Other satellites in this generation include ViaSat 1 and Jupiter 1. KA-SAT, with a launch mass of 6,150 kg, is totally dedicated to Ka-band broadband access. It is also the first Ka-band HTS in Europe. It was manufactured by EADS Astrium (now Airbus Defence and Space) based on its Eurostar 3000 platform and launched on 26 December 2010 on a Proton-M/Briz-M launcher to the orbital position of 9°E. Figure 7.1 shows the satellite, which has an estimated maneuver lifetime of 16 years.

The payload requires 11 kW of the total of 14 kW of DC power available from the power system to supply 56 TWTAs rated at a saturated power of 130-W RF. It includes 82 spot beams each covering an area of approximately 250 kilometers in diameter and providing approximately 900 Mbps of capacity [5, 6]. The service area covers Europe and the Mediterranean Basin and parts of the Middle East, as depicted in Figure 7.2.

Each user beam has an available spectrum for the terminals of 237 MHz on the uplink and another 237 MHz on the downlink [7]. Single polarization is used on each cell for the uplink and downlink, while the uplink polarization is orthogonal to that of the downlink. The user links exploit the Exclusive

Figure 7.1   An image of the KA-SAT spacecraft in the test range. (*Source:* Airbus Defence and Space. Reprinted with permission.)

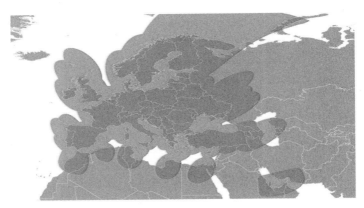

Figure 7.2 The user service area of KA-SAT.

Bands in Ka-band as defined by the Radio Regulations of the ITU [8] and as supported by the CEPT [9] (i.e., 29.5–30.0 GHz on the uplink and 19.7–20.2 GHz on the downlink). This is an administrative asset in terms of the licensing requirements for the terminals operating in these frequencies within the participating CEPT countries. However, it also limits the capacity.

KA-SAT employs a four-color scheme so that 500-MHz segments available in the Exclusive Bands are split into four resources on both polarizations. This gives a frequency reuse of 20 over the service area. This is an essential element of HTSs, since the system bandwidth becomes 20 times the available spectrum in order to achieve a total system capacity that exceeds 90 Gbps [7].

The gateways utilize 1.25 GHz on both polarizations of the remaining spectrum in the Ka-Band, namely 17.7–19.7 GHz and 27.5–29.5 GHz. To fully match the user capacity, eight operational gateways are required with each gateway serving 10 user beams. However, KA-SAT was designed to operate with eight active gateways of 10 available to cater for propagation effects and gateway downtime.

KA-SAT can deliver Internet connectivity to more than one million homes, at speeds comparable to Asymmetric Digital Subscriber Line (ADSL). KA-SAT started with the TooWay service employing ViaSat Surfbeam technology. (See Figure 7.3.) It provides an always-on service delivering speeds of up to 50-Mbps download and 6-Mbps upload. Professional requirements for higher volume and bandwidth are possible with speeds of up to 50-Mbps download and 10-Mbps upload [10].

KA-SAT also enables cost-effective satellite news gathering (SNG) using lightweight, transportable uplink antennas, regional and local TV, and corporate TV networks.

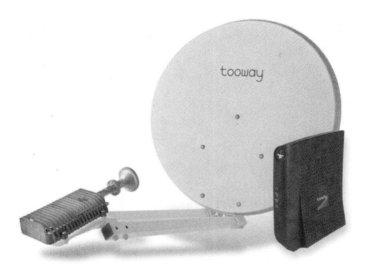

Figure 7.3 A KA-SAT terminal.

7.2.2   Eutelsat 36C/Express AMU1

Eutelsat 36 C/Express AMU1 is a satellite owned by the Russian Satellite Communication Company (RSCC), but it was defined together with Eutelsat, which leases capacity on the satellite. The satellite is part of the 36° East neighborhood with another satellite from Eutelsat. The position supports TV broadcasting for TV broadcasting in the Commonwealth of Independent States, and Sub-Saharan Africa. Manufactured by Airbus Defence and Space on its Eurostar-3000 platform, the satellite was launched on a Proton-M Briz-M (Ph.3) launcher on 24 December 2015. At lift-off, the mass of the satellite was 5,700 kg. The satellite can deliver 15 kW of power and has a lifetime of 15 years. Figure 7.4 depicts the satellite, which can operate up to a total of 70 TWTA.

Besides the broadcast capacity in the Ku-band, Eutelsat 36 C/Express AMU1 includes an HTS payload in the Ka-band covering the western part of the Russian Federation and part of Western Siberia. The satellite employs 10 Ka-band TWTAs, has 18 user spot beams (as shown in Figure 7.5), and one active gateway to deliver a total of 11-Gbps capacity [11].

For the ground segment, Eutelsat uses a Gilat's SkyEdge II-c hub with X-Architecture and SkyEdge II-c small user terminals to deliver broadband [11]. The hub is installed at the Dubna satellite center, which is situated near

Figure 7.4   Picture of the spacecraft Eutelsat 36 C/Express AMU 1 in the test range. (*Source:* Airbus Defence and Space. Reprinted with permission.)

Moscow and operated by RSCC. The service, marketed as Konnect Russia, offers consumer and professional subscribers Internet access with speeds of up to 40 Mbps downstream and 12 Mbps upstream [12, 13]. Following the Express-AM6 anomaly, a Hughes Jupiter gateway is also used on Eutelsat 36 C/Express AMU 1 to recover most of Express-AM6 traffic [14].

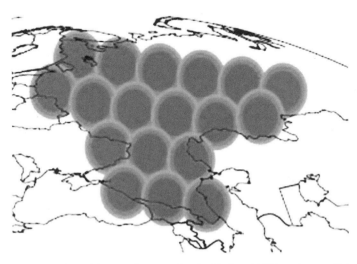

Figure 7.5   Illustration of the service area for the Ka-band HTS system on Eutelsat 36 C/Express AMU 1.

7.2.3 Eutelsat 65WA

Eutelsat 65 West A was manufactured by SSL (now Maxar) on the SSL-1300 platform. It was launched on 9 March 2016 on the Ariane-5 ECA launcher with a wet mass of 6,564 kg. Figure 7.6 shows a picture of the spacecraft in the test range. It is the first Eutelsat satellite on the orbital position 65° West. It has a design lifetime of 15 years, and the satellite can deliver 16.7 kW at the end of life (EOL).

Eutelsat 65 West A is a triband satellite designed to target video and broadband markets across Latin America reaching 18 million homes. Up to 66 TWTAs can be used to power the three payloads. The Ku-band payload supports DTH reception across Brazil and facilitates corporate connectivity in Central America, the Caribbean, and the Andean region. The satellite also

Figure 7.6 A picture of the Eutelsat 65 West A spacecraft in the test range. (*Source:* Maxar Technologies. Reprinted with permission.)

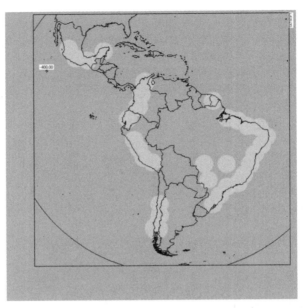

Figure 7.7   The service area of the Ka-band HTS system on Eutelsat 65 West A.

features a transatlantic C-band coverage for cross-continental video contribution and distribution.

The Ka-band HTS system uses 24 Ka-band TWTAs to support 24 spots from a possible 32 over a significant area as illustrated in Figure 7.7. Providing a total (forward and return) capacity of 35 Gbps, the mission focuses on the major metropolitan urban and suburban areas of Brazil and the coastlines of Central and South America, allowing Eutelsat to meet the growing broadband demand in Latin America [11].

The entire Brazilian capacity of 24 Gbps over 16 spots is commercialized by Hughes Network Systems (HNS) do Brazil, an EchoStar company [15, 16]. The system over Brazil covers 85% of the population and employs the high-throughput JUPITER technology from HNS. The remaining capacity of the HTS capacity is commercialized by Stargroup to provide services in Mexico, Colombia, and Peru [15, 17].

7.2.4   Eutelsat 172B

EUTELSAT 172B is a high-capacity satellite delivering capacity for fast-growing applications in Asia Pacific, including in-flight and maritime connectivity, cellular backhaul, corporate networks, video, and government services.

The EUTELSAT 172B satellite is a triple mission satellite with the following three distinct payloads:

- A trans-Pacific C-band payload with 14 transponders delivering increased power and broader coverage to enhance the service previously provided via Eutelsat 172A as well as reach new growth markets in South East Asia;
- A Ku-band payload with 36 transponders, which doubles capacity at 172° East and connects five improved service areas: the North Pacific, North-East Asia, the South-East Pacific, the South-West Pacific, and the South Pacific;
- An innovative high-throughput Ku-band payload designed for in-flight broadband with multiple spot beams optimized to serve densely used Asian and trans-Pacific flight paths.

The satellite was manufactured by Airbus Defence and Space on its Eurostar E3000 EOR platform. This is the version of the platform employing electrical propulsion and its first application. Figure 7.8 provides an illustration of the spacecraft. An Ariane 5 ECA was launcher that lifted-off on 1 June 2017. The launch mass of the satellite was 3,551 kg, and it can deliver 13-kW EOL with a design lifetime of 15 years.

The HTS payload operates in the Ku-band and features a multispot coverage optimized for IFC applications to deliver 1.8 Gbps of capacity over the densely used Asian and trans-Pacific flight paths as illustrated in Figure

Figure 7.8 A rendering of the Eutelsat 172B spacecraft. (*Source:* Airbus Defence and Space. Reprinted with permission.)

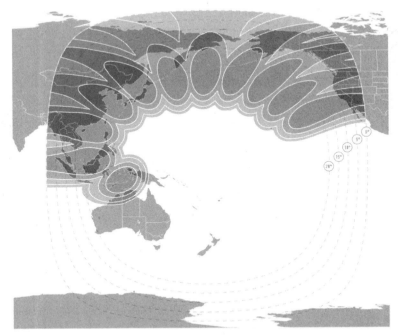

Figure 7.9   The service area of the Ku-band HTS payload on Eutelsat 172B.

7.9. The gateways operate in the Ka-band. The use of a multiport amplifier ensures the flexible assignment of the forward capacity among the 11 elliptical beams so that the limit is the total power available to all beams while allowing great flexibility in the apportionment of the power amongst the beams. This translates to a limit in the total capacity with less constraints on the capacity per beam, making the system more efficient and operationally versatile.

The HTS capacity is leased to Panasonic Avionics and China Unicom for IFC over the flights in the region [18, 19].

7.2.5   Konnect

Konnect, initially called Broadband for Africa (BB4A), was constructed by Thales Alenia Space using the new all electric Spacebus-Neo 100 platform. Figure 7.10 shows the spacecraft in the anechoic chamber. It lifted off on 16 January 2020 on Ariane 5 ECA+ with a mass of 3,619 kg. Its design lifetime is 15 years with a payload power of 7 kW. At the time of writing, the satellite is still in its electrical orbit raising phase. Konnect is Eutelsat's contribution to

Figure 7.10 Photo of Konnect in the anechoic chamber. (*Source:* Thales Alenia Space. Reprinted with permission.)

help to fight against the digital divide by bringing broadband Internet across 40 countries in Africa and 15 countries across Europe.

The satellite's main mission is Konnect Africa, but it also has the possibility of supporting Konnect Europe with significant flexibility for the process of transferring capacity from Europe to Africa. For Konnect Africa, the satellite uses 38 Ka-band TWTAs to deliver 75 Gbps of capacity across a network of 65 spotbeams, which together provide quasi-complete coverage of Sub-Saharan Africa [20]. The service area is shown in Figure 7.11. An additional 28 beams are for Konnect Europe. Six active gateways are required to support this capacity. The satellite will address direct-to-user consumer and

Figure 7.11   The service area of Konnect Africa.

enterprise broadband services using dishes from approximately 75 cm. It will also be used for community networks connected to Wi-Fi hotspots, mobile phone backhauling, and rural connectivity.

Konnect matches the available system capacity to the expected geographical capacity demand. It also provides flexible allocation of spectral resources between the spots to allow for distribution of capacity as commercially required over the lifetime of the satellite. The service is now part of the Eutelsat Konnect initiative, which aims to boost social and economic development, providing reliable and fast connectivity, anywhere.

Konnect Africa uses the General Dynamics SATCOM Technologies' solution with seven 9-m antennas for the gateways and the JUPITER™ ground network system from HNS, including baseband equipment and new generation user terminals with an antenna of 75 cm [21]. A memorandum of understanding has been signed with Schoolap and Flash Services to connect several thousands of schools across the Democratic Republic of Congo (DRC) and provide them with high-speed Internet connectivity This allows them to access to an official digital platform for educational content and teaching materials [22].

7.2.6 Future HTSs

The Konnect Very High–Throughput Satellite (VHTS), the next satellite to be launched by Eutelsat, is currently under production at Thales Alenia Space on its Spacebus Neo platform for an expected entry to service in 2022. As the name implies, it is intended to significantly exceed the capacity of current Eutelsat HTS. Konnect VHTS is designed to deliver 500 Mbps in the Ka-band [23]. The payload includes a transparent digital processor to enhance the on-board allocation of spectral resources and the phased introduction of gateways following the traffic take up. It will employ V-band for the gateways. The launch mass is estimated to be 6,300 kg.

Konnect VHTS already has firm multiyear distribution commitments from Orange and Thales. A retail partnership was signed with Orange to address the fixed-broadband market in European countries, and a distribution partnership was inked with Thales to serve notably the government connectivity services market.

7.3 The Viasat Satellites

Viasat started the broadband satellite business with capacity on Anik F2 to deliver a service with a download speed of 1.5 Mbps. In 2004, Wildbue Communications launched its satellite Internet service over the 48 contiguous states of the United States with capacity on the Canadian satellite Anik F2. In 2006, WildBlue 1 was launched to cover North America, and the service was subsequently upgraded with the new asset. On 15 December 2009, Viasat bought WildBlue with its capacity on Anik F2 and the WildBlue 1 satellites to provide a service with download speeds of 1.5 Mbps and 3 Mbps, respectively.

Viasat states that it is on a mission to solve one of the most challenging communications problems: making the internet accessible and affordable for all. It recognizes that the economic impact of the Internet on society has not yet been fully realized, as there are still many rural and urban communities without access to high-quality broadband. Viasat believes satellite to be a critical part of the broadband ecosystem as the most economical and sustainable way to serve billions of people with fast, high-value connectivity services that promote digital inclusion, transform how people communicate, and encourage new economic development opportunities through advancements in education, healthcare, agriculture, and more.

It is interesting to note that the ViaSat fleet, although small, is dedicated to Ka-band broadband connectivity.

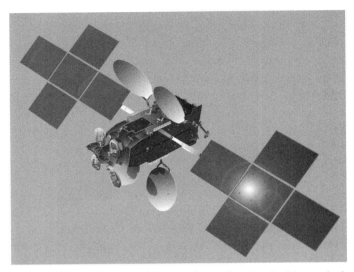

Figure 7.12   A rendering of ViaSat 1. (*Source:* Viasat. Reprinted with permission.)

The ViaSat 1 satellite was manufactured by SSL (now Maxar) on the 1300 platform and launched on 19 October 2011 aboard a Proton-M Briz-M (Ph.3) launcher to be positioned at 115.1°W. At launch, its mass was 6,740 kg; it has a design lifetime of 15 years. Figure 7.12 provides a rendering of the spacecraft. It held the Guinness record for the world's highest capacity communications satellite with download of up to 12 Mbps at the time and a total capacity in excess of 140 Gbps at the time of its launch [24, 25]. A download speed of up to 50 Mbps using newer-generation user terminals is currently claimed.

As can be seen in Figure 7.13, ViaSat 1 delivers its capacity over 72 Ka-band spot beams, 63 over the United States (eastern and western states, Alaska and Hawaii), and nine over Canada. The Canadian beams are owned by satellite operator Telesat and leased to Xplornet Communications International for the Xplornet broadband service to consumers in rural Canada. The U.S. region is roughly split into three areas: the west area, the east area, and the central area. The user service area is basically the west and the east. This offers the opportunity of locating the gateways in the central area so that frequency reuse of the Ka-band between the gateway and the user cells is possible, providing considerable spectrum for both the cells and the gateways and increasing the total system capacity. The U.S. beams provide fast Internet access originally branded as Exede, ViaSat's satellite Internet service at the time. ViaSat 1 shares the same generation of ground segment as KA-SAT, the

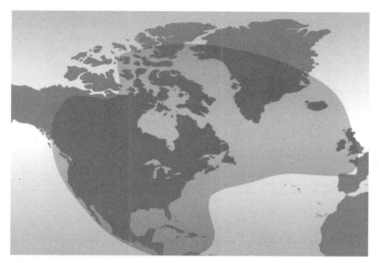

Figure 7.13 Coverage of ViaSat 1, showing the user spot beams and gateway locations.

SurfBeam® 2 satellite networking system. ViaSat 1 delivers DSL and wireless broadband services [26].

Viasat 1 was followed by Viasat 2, which was constructed by Boeing on its BSS 702HP platform. It was launched on 1 June 2017 using the Ariane 5 ECA launcher. At lift-off, the satellite mass was 6,418 kg, and it was positioned at 69.9°W. The satellite, which has a design lifetime of 14 years, delivers 16.1 kW at EOL. Figure 7.14 provides a rendering of the spacecraft. The satellite was intended to have at least 300 Gbps of capacity, but due to an antenna anomaly the satellite delivered 260 Gbps [27] in the Ka-band. After a seven-month period of orbit raising, it was revealed that two of the four spacecraft antennas had a malfunction that reduced their efficiency.

The number of beams on ViaSat 2 has not been publicly disclosed, but it is known that there are two beam sizes [28], probably to cater for the dense areas of high capacity and the sparse areas of the expanded coverage. Viasat published only an envelope coverage, as shown in Figure 7.15, with an extended footprint across North America, Central America, the Caribbean basin, and the Atlantic Ocean between North America and Europe. It is intended to deliver broadband services with download speeds of up to 100 Mbps to markets that include residential, in-flight, maritime, emergency relief and response, oil and gas, and government.

Figure 7.14 A rendering of the ViaSat 2 spacecraft. (*Source:* Viasat and Boeing Satellite Systems. Reprinted with permission.)

The ViaSat 3 fleet of satellites is currently under production with the BSS 702HP coming from Boeing and Viasat producing the payloads. Three satellites are planned with launches scheduled for 2022 for three global regions. Each satellite is expected to deliver 1 Tbps of Ka-band capacity, and the launch mass is expected to 6,400 kg. The three targeted regions are the Americas, Europe, and the Asia-Pacific region.

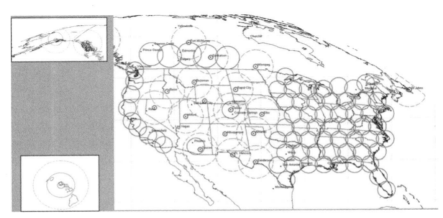

Figure 7.15 The envelope service area of Viasat 2.

7.4 The Jupiter Satellites

In 2007, Hughes launched the Spaceway 3 next-generation regenerative Ka-band satellite system. It employed advanced techniques including on-board fast packet switching and routing and spot beamforming. This was beyond the performance of what most current digital processor offer. Spaceway 3, launched in August 2007, was used to deliver a wide range of HughesNet IP services to enterprise, government, and consumer/small business customers in North America. The satellite, which deliver 10 Gbps overall capacity, ushered in a new world of bandwidth-on-demand satellite services, with true peer-to-peer, single-hop networking of high-performance ground terminals. In 2008 the network supported downloads of up to 2 Mbps; this increased to 5 Mbps in 2013 [29].

On 14 February 2011, EchoStar acquire Hughes Communications in a deal valued at $1.3 billion, and HNS became a wholly owned subsidiary of EchoStar. The spacecraft operations are performed by Echostar, while the network operations and management are in the hands of Hughes. This could explain why there are multiple names for the satellite stemming from the parent company and from Hughes. Residential broadband connectivity is the major activity of Hughes.

Today, Hughes has over 1.3 million high-speed Internet subscribers and offers services in the United States, Canada, and Brazil over Echostar satellites and leased capacity.

Jupiter 1, also known as Echostar XVII or Echostar 17, was manufactured by SSL (now Maxar) using its 1300 platform and a power capacity of 16.1 kW. The satellite has a design lifetime of 15 years. Figure 7.16 shows an image of the spacecraft in the test range. It was launched on 5 July 2012 using the Ariane 5 ECA launch vehicle. Its mass at lift-off was 6,100 kg, and the orbital positions is 107.1° West [30].

The satellite delivers 120 Gbps of Ka-band capacity over 60 beams [31] to support markets including consumer, enterprise, aeronautical, cell backhaul, and community WiFi. Figure 7.17 shows the coverage of the satellite, which started operations with the Hughes Gen4 Internet service using the Hughes JUPITER™ System as its satellite networking platform. In 2017, Jupiter 1 upgraded to Hughes Gen5 to provide downloads of up to 25 Mbps and uploads of 3 Mbps [32].

Jupiter 2, also referred to as Echostar 19 or Echostar XIX, was manufactured by SSL (now Maxar) employing its 1300 platform. The platform generates 16.4 kW, and the satellite has a design lifetime of 15 years. At lift-off, its mass was 6,637 kg, and it was launched on 18 December 2016 using the

Figure 7.16   The Jupiter 1 spacecraft in the test range. (*Source:* Maxar Technologies. Reprinted with permission.)

Atlas 5(431) launcher for deployment at 97.1° West. The satellite is illustrated in Figure 7.18.

The satellite delivers 220 Gbps of Ka-band capacity [33] over 139 user beams and using 22 gateways [34, 35]. Figure 7.19 shows the coverage of Jupiter 2. Seventeen gateways are in CONUS, two are in Canada, and three

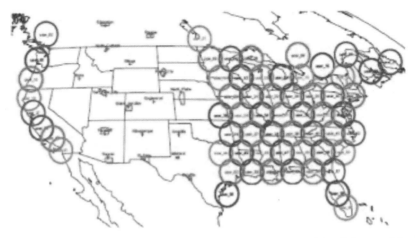

Figure 7.17   Coverage of the Jupiter 1 (also known as Echostar XVII or Echostar 17) [31].

Figure 7.18 Jupiter 2 (Echostar 19 or Echostar XIX) in the test range. (*Source:* Maxar Technologies. Reprinted with permission.)

are in Mexico. Out of the 139 user beams, 90 are in CONUS, two in Alaska, 16 in Canada, 24 in Mexico, five in Central/South America, one in Puerto Rico, and one in Cuba. Jupiter 2 supports the Hughes Gen5 together with other satellites. The system incorporates the system-on-a-chip (SoC) technology, incorporating the latest DVB-S2X air interface standard, wideband 250-MHz carriers operating at more than 1 Gbps, and throughputs of over 100 Mbps on every terminal.

Jupiter 3 (also known as Echostar 24 or Echostar XXIV) is currently under production at Maxar employing the 1300 platform. It will also operate in the Ka-band to deliver 500 Gbps over the Americas when launched in 2021

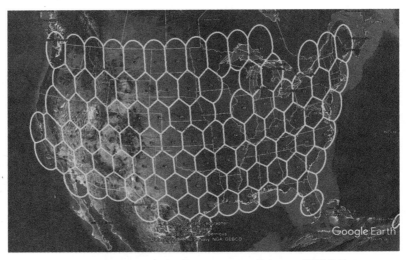

Figure 7.19   The coverage of Jupiter 2 (Echostar 19 or Echostar XIX) [36].

[37]. Hughes has called it an ultra-high-density satellite (UHDS) because of the capacity it delivers over the region.

7.5   Global Xpress

INMARSAT has its origins in the intergovernmental organization that was set up 1979 at the request of the International Maritime Organization, a United Nations maritime body. In the 1980s, the convention governing INMAR-SAT was amended to include improvements to aeronautical communications, notably for public safety in coordination with the International Civil Aviation Organization. In April 1999, the operational part of the original organization became a private U.K. company known as Inmarsat, Ltd. Until the 2010s, Inmarsat provided mobile global services in the L-band mostly for the maritime and aeronautical community.

Inmarsat started Ka-band HTS services with its fifth-generation satellites, namely Inmarsat 5 F1 to F5 or Global Xpress GX1 to GX5. Table 7.1 provides these spacecraft's details.

GX1 to GX4 share the same design from Boeing and include 89 fixed spot beams over Earth as seen by a given satellite and six steerable antennas. Figure 7.20 illustrates the GX spacecraft. Out of the 89 fixed beams, 72 can be active [38]. Figure 7.21 depicts the coverage of the fleet. This means that each fixed beam has a diameter of about 1.8°, assuming an 75% coverage of

Table 7.1
Details of the Inmarsat GX1 to GX5 Satellites [38–44]

Satellite	GX1	GX2	GX3	GX4	GX5
Satellite Supplier	Boeing	Boeing	Boeing	Boeing	TAS
Platform	BSS 702HP	BSS 702HP	BSS 702HP	BSS 702HP	Spacebus 4000B2
EOL Power	13.8 kW	13.8 kW	13.8 kW	13.8 kW	
Payload Power					6.8 kW
Launch Mass	6,100 kg	6,100 kg	6,100 kg	6,100 kg	4,007 kg
Launch Vehicle	Proton Briz M	Proton Briz M	Proton Briz M	Falcon 9 v1.2	Ariane 5ECA
Launch Date	8 Dec 13	2 Feb 15	28 Aug 15	15 May 17	26 Nov 19
Orbital Position	62.6° West	55° West	179.6° East	56.5° East	11° East
Lifetime	15 years	15 years	15 years	15 years	16 years

the visible Earth. On each spot beam, 2×40 MHz of bandwidth is available for the forward and the return, respectively. The steerable beams have channels of 125 MHz for both directions. Inmarsat does not provide figures for system capacities. However, the total system capacity of each satellite is estimated to be about 7.5 Gbps assuming the utilization of 60-cm terminals. Please note that this estimate is based on several assumptions. Although the estimate may appear low, one needs to consider that this is a global system; accordingly, the beamwidths are relatively large, and the terminals are mobile, so performance is compromised for size.

An interesting feature of the Inmarsat global requirements is that the satellites of the fleet are practically identical since the footprint from the satellite for the different orbital positions is similar. Having a fleet of identical satellites is a rare opportunity for satellite operators.

GX5 was designed and built by Thales Alenia Space to provide more capacity over Europe and the Middle East through 72 Ka-band spot beams. It is one of the first HTS satellites to use the V-band for feeder link. Unlike GX1 to GX4, GX5 includes a digital processor [44]. GX5 departs from the traditional Inmarsat satellites since this satellite is more regional than global.

The next satellites to be launched by Inmarsat are Inmarsat 6 F1 and F2, also referred to as GX6A and GX6B. They are being manufactured by

Figure 7.20   A rendering of an Global Xpress spacecraft. (*Source:* Inmarsat. Reprinted with permission.)

Airbus Defence and Space and will utilize the all-electric E3000 platform [45]. They are scheduled for launch in 2021. They are hybrid satellites with both an L-band payload and a smaller Ka-band payload. The satellite embarks a digital processor and nine Ka-band steerable antennas that can be pointed as and where required on the visible Earth and with the required capacity allocation.

The seventh generation of Inmarsat satellite Inmarsat 7 F1, F2, and F3, or GX7, GX8, and GX9 will be based on Airbus' new standard and fully reconfigurable satellite product line OneSat. Featuring on-board processing and active antennas, the three Ka-band spacecraft will be able to optimize their coverage and allocate RF power and spectral resources as required to best meet the market requirements for mobility [46].

The GX fleet builds on Inmarsat's mission to provide mobile services particularly to the maritime, government, and aeronautical sectors. In 2019, the three sectors represented 35%, 29%, and 21% of the revenue, respectively,

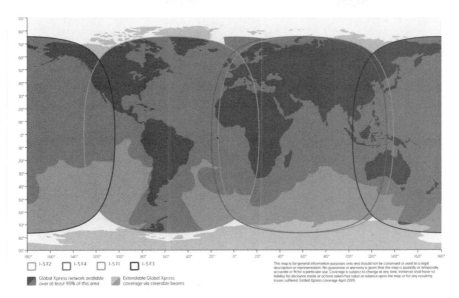

Figure 7.21 The coverage of the Inmarsat satellites Global Xpress GX1 to GX4.

based on the first half figures [47]. The GX services that started in 2015 and for the first half of 2019, the related revenues represented 24% of the total revenues. The GX fleet supports communications to over 7,000 vessels, 600 business jets, and 27 commercial airlines.

7.6 Sky Muster

Sky Muster is an HTS system that is operated by NBN Co., an Australian government–owned corporation tasked to design, build, and operate National Broadband Network for Australia as a monopoly wholesale broadband provider. It was established in 2009 with the purpose of connecting Australia and bridging the digital divide to some 400,000 homes and businesses.

NBN is structured as a wholesale-only, open-access broadband network. NBN provides services on its local access network on equivalent terms to retail phone and Internet providers to provision for end-user needs. This is intended to level the playing field in Australian telecommunications, creating real and vibrant competition within the industry and providing choice for consumers.

There are two satellites in the constellation, namely Sky Muster I and Sky Muster II (formerly called NBN Co 1A and 1B). The two satellites are

Figure 7.22   Picture of Sky Muster spacecraft in the test range. (*Source:* Maxar Technologies. Reprinted with permission.)

very similar, and both have been manufactured by SSL/Maxar using its 1300 platform and have a design lifetime of 15 years. The EOL power delivered by the platform is 16.4 kW. At lift-off each satellite had a mass of 6,440 and 6,405 kg, respectively [48, 49]. Sky Muster I was launched on 1 October 2015 and was deployed at 140° East, while Sky Muster II was launched at 145° East. Figure 7.22 shows one of the satellites in the test range.

Each satellite includes 101 spot beams covering Australia and some of the remote islands, namely Norfolk, Christmas, Macquaire, and Cocos Islands. The satellites operate in the Ka-band, and the service employs Ka-band technology and terminals. Each beam is used for the forward and the return links. Sky Muster is a prime example of the flexibility in the design of an HTS satellite, covering small remote areas, large expanses of sparsely populated area in central Australia, and the relatively dense belts on Western and Easter Australia. Figure 7.24 illustrates its coverage. Small beams are used either when the area (island) to be covered is small or when higher capacity density is required. Large beams ensure a ubiquitous coverage even when the population density is low. For example, in the Northern Territory, the density is just 0.2 people per square kilometer. The most populated is the Australian Capital Territory with 174 habitants per square kilometer.

The constellation delivers 180 Gbps of capacity and with downloads between 12 and 25 Mbps and uploads of 1–5 Mbps with terminals of aperture

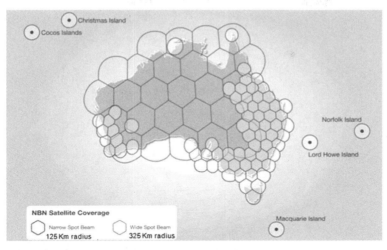

Figure 7.23 Coverage of the Sky Muster constellation.

diameter of 80 or 120 cm [50, 51]; 94% of the users have the smaller aperture terminals. The ground segment, provided by Viasat, includes nine active gateways with diameters of 13.5m [52]. The two-satellite constellation allows for the optimization of the coverage for specific users but also increases the resilience of the system through redundancy.

7.7 The SES HTS Satellites

SES is one of the world's leading satellite owners and operators with over 70 satellites in orbit. SES started satellite operations with the Astra satellites broadcasting TV over Europe. SES gas grown through acquisitions and partnerships to operate over satellites with the names of AMC, Ciel, NSS, QuetzSat, YahSat, and of course SES. The latest new name is O3b. Most of SES's satellites are in the GEO orbit, but it also operates the O3B constellation of 20 satellites in the equatorial MEO orbit at an altitude of 8,000 km.

SES operates three HTS satellites: SES 12, SES 14, and SES 15 form the current SES fleet. It is interesting to note that SES started HTS operations in 2018, considerably later than the introduction of KA-SAT, ViaSat 1, and Jupiter 1 in the 2011–2012 period. Unlike other HTS operators, SES delivers GEO HTS services entirely in the Ku-band. This is probably because the target market was aeronautical IFC, which was more mature in the Ku-band.

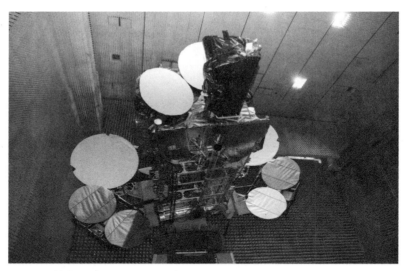

Figure 7.24 SES 12 in the test range. (*Source:* Airbus Defence and Space. Reprinted with permission.)

With the launch of SES 17, SES will enter the league of the mainstream HTSs in the Ka-band.

SES 12 was built by Airbus Defence and Space on its Eurostar E3000 platform with EOR with solar arrays generating 19 kW of power. It is designed for a lifetime of 15 years. It was launched on June 4, 2018 as the payload of 9 version 1.2 with mass of 5,384 kg. Its nominal operational position will be 95°E [53, 54]. Figure 7.24 shows the spacecraft in the test range.

Beside the legacy broadcasting mission, SES 12 also has a flexible multibeam payload for providing broadband services covering a large expanse from Africa to Russia, Japan, and Australia. SES-12 operates a total of 76 active TWTs with 72 HTS beams and is equipped with eight antennas. The user beams operate in the Ku-band, and the gateways operate in the Ka-band. The satellite is equipped with a digital transparent processor that facilitates spectrum management on-board the spacecraft to increase the flexibility and efficiency. The Ku-band HTS system targets mostly the aviation, maritime, and cellular backhaul services. Figure 7.25 illustrates the Ku-band envelope coverage for SES 12. The satellite reached the intended orbital location and went operational on February 26, 2019 [55, 56].

SES 14 was constructed by Airbus Defence and Space on its Eurostar 3000 all-electric platform with an electrical power capability of 16 kW. The lift-off mass was 4,423 kg, and it was launched on Ariane 5 ECA on January

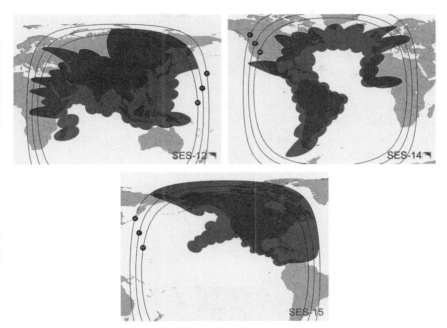

Figure 7.25 Envelope coverage of the Ku-band HTS system on SES 12, SES 14, and SES 15.

Figure 7.26 Picture of SES 14 in the test range. (*Source:* Airbus Defence and Space.)

25, 2018 for deployment at 47.5° West. The design lifetime of the satellite is 15 years. Figure 7.26 shows the satellite under testing.

Like SES 12, the mission of SES 14 is hybrid, carrying legacy C-band and Ku-band capacity but also an HTS capacity (i.e., replacement and growth capacity). The users operate in the Ku-band, while the gateways are in the Ka-band. It also carries a digital transparent processor to ease on-board spectral management. As shown in the SES 14 coverage in Figure 7.25, the target coverage is the Americas and the North Atlantic where it aims to render services to the aeronautical, maritime, and cellular backhaul markets. The concept of the Ku-band HTS system is similar to that of SES 12, although smaller. After completion of the orbit raising phase and testing, the satellite became operational on September 4, 2018 [57].

SES 15 was manufactured by Boeing Satellite Systems on its all-electric BSS 702 SP platform with an 8-kW payload and a design lifetime of 15 years. It was launched by Soyuz-ST-A Fregat-M on May 18, 2017 with a lift-off mass

Figure 7.27 The spacecraft SES 15. (*Source:* SES and Boeing Satellite Systems. Reprinted with permission.)

of 2,302 kg [58]. The operational orbital position is 129° West. Figure 7.27 pictures SES 15.

SES 15, like SES 12 and 14, is a hybrid satellite with legacy capacity and an HTS payload with users in the Ku-band and gateways in the Ka-band. SES 15 covers North America, Latin America, and the Caribbean as shown in the coverage in Figure 7.25 for SES 15. The main targeted market is aeronautical together with government, VSAT networks, and maritime [59].

SES 17 is currently under production at Thales Alenia Space with a design based on the all-electric Spacebus NEO. The design lifetime of the satellite will be 15 years. SES 17 is scheduled for launch on Ariane [60]. It will have a digital payload that requires 15 kW. The payload, which will include close to 200 spot beams of different sizes, particularly suited for the aeronautical IFC providing optimal coverage for the corridors in North America, the Caribbean, Mexico, Latin America, and the Atlantic Ocean. SES has secured a long-term commercial agreement with Thales to offer FlytLIVE, a new inflight connectivity service over the Americas and the Atlantic Ocean region [61]. SES will operate the satellite infrastructure for FlytLIVE, as well as the complementary ground network. The satellite is expected to be launched in 2021.

7.8 The International Telecommunications Satellite Organization (Intelsat) Epic^{NG} Satellites

INTELSAT is longstanding satellite operator that started as an intergovernmental organization in 1964. INTELSAT was originally formed to own and manage a constellation of communications satellites providing international broadcast services. In 1965, INTELSAT launched the world's first commercial communications satellite, Intelsat 1, or as it is more commonly known as Early Bird. In 2001 in line with other satellite IGOs, INTELSAT became a private company, Intelsat, Ltd., based in Luxembourg. It has had the largest fleet of satellites for a long time and currently operates 49 satellites. However, Intelsat filed for a Chapter 11 bankruptcy in U.S. courts on May 13, 2020.

In 2012 Intelsat announced the Epic^{NG} satellite platform. Intelsat Epic^{NG} is based on an open architecture and is engineered for backward compatibility, allowing network operators to realize high-throughput performance utilizing their existing hardware and network infrastructure. The satellite flexibility allows network operators to offer their end users a customized network with the required topology, hardware, and service characteristics. This is an important statement when systems are compared.

In line with the Intelsat philosophy, Intelsat Epic^{NG} is intended to be a worldwide system providing broadband connectivity, enabling delivery of a greater variety of solutions to more customers with improved economics. The global vision of Intelsat is seen in the coverage of the Epic^{NG} satellite as illustrated in Figure 7.28.

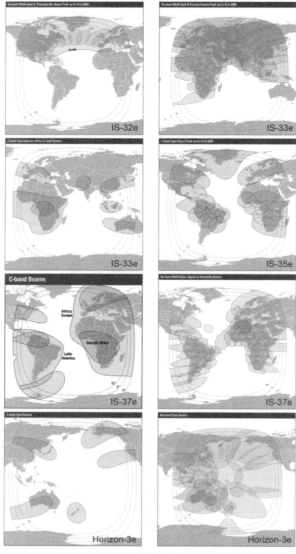

Figure 7.28   Coverage of the Intelsat Epic^{NG} fleet of five satellites.

Table 7.2
Details of the Launched Intelsat Epic[NG] Satellites [62–67]

Satellite	Intelsat-29e	Intelsat-32e	Intelsat-33e	Intelsat-35e	Intelsat-37e	Horizon-3e
Satellite Supplier	Boeing	Airbus	Boeing	Boeing	Boeing	Boeing
Platform	BSS 702MP	E3000	BSS 702MP	BSS 702MP	BSS 702MP	BSS 702MP
EOL Power	15.8 kW	16.0 kW	13.0 kW	12.0 kW	14.0 kW	18.5 kW
Launch Mass	6,552 kg	6,000 kg	6,600 kg	6,761 kg	6,438 kg	6,441 kg
Launch Vehicle	Ariane 5ECA	Ariane 5ECA	Ariane 5ECA	Falcon-9 v1.2	Ariane 5ECA	Ariane 5ECA
Launch Date	27 Jan 16	14 Feb 17	24 Aug 16	05 Jul 17	27 Sep 17	25 Sep 18
Orbital Position	50° West	43° West	60° East	34.5° West	18° West	169° East
Lifetime	15 years	19 years	15 years	15 years	15 years	15 years

Intelsat launched six Epic[NG] satellites as listed in Table 7.2 and currently operates five of them.

Intelsat 29e (also referred to as IS-29e) reached its test position at 49.7° West, deployed its antennas and solar panels, and successfully completed the payload in-orbit testing. However, on 7 April 2019, the propulsion system of Intelsat 29e developed a fuel leak, and Intelsat issued a statement declaring the satellite a total loss on 18 April 2019. Figure 7.29 shows the Intelsat 37e.

Intelsat 32e shares the broadcasting Sky Brazil-1 payload. Intelsat 33e arrived on station three months late due to a thruster malfunction. Horizons 3e is owned jointly by Intelsat and SKY Perfect JSAT as part of a joint venture.

Although there are similarities between the Intelsat Epic[NG] satellites and the traditional HTS satellites, their mission and design are different. The main objective is an enhanced overlay to the space segment they already provide. They do not attempt to reduce the total cost through a closed system and thus close the door for opportunities that lie outside the closed system. They provide an enhanced air interface through spacecraft spot beam technology as HTS systems do. Unlike most HTS systems they operate mostly in the Ku-band and C-band with cross-strapping, and in the case of Intelsat 35e, three contoured Ku-band beams are available over Europe and Africa; the Caribbean and Europe are also available.

Figure 7.29   A rendering of the Intelsat 37e (IS-37e) spacecraft. (*Source:* Intelsat and Boeing Satellite Systems.)

In the Boeing satellites, the connectivity between the beams is enhanced through the digital processor [68–70], and the RF power is efficiently distributed among the beams through multiport amplifiers [71]. The Epic^{NG} payloads support star, mesh, and loop back network topologies.

The next Epic^{NG} satellite to be launched, Intelsat 40e, will be manufactured by Maxar Technologies on its 1300 platform. It is scheduled for launch on Falcon 9 in 2022 [72].

References

[1]   Viasat, "Status Update for ViaSat-2—our Newest Satellite," August 23, 2017, https://corpblog.viasat.com/status-update-for-viasat-2-our-newest-satellite-2/.

[2]   Thaicom, *IPStar Project, Fact Sheet Q2/2014*, https://www.thaicom.net/wp-content/uploads/2019/07/20140901-factsheet-2q2014-ipstar-en.pdf.

[3]   Thaicom, *IPStar Project, Fact Sheet Q2/2013*, https://www.thaicom.net/wp-content/uploads/2019/07/Factsheet-2Q2013-619-EN.pdf.

[4]   Eutelsat, *Third Quarter and Nine Month 2019–20 Revenues*, https://www.eutelsat.com/files/PDF/investors/2019-20/Eutelsat_Communications_Q3_2019-20_PR.pdf.

[5]   Fenech, H., et al., "The KA-SAT System," *15th Ka and Broadband Conference*, Cagliari, Italy, Istituto Internazionale delle Communicazioni, September 23–25, 2009.

[6]   Fenech, H., et al., "KA-SAT and the Way Forward," *Ka and Broadband Communications, Navigation and Earth Observation Conference*, Palermo, Italy, Istituto Internazionale delle Communicazioni, October 3–5, 2011.

[7]   Fenech, H., "KA-SAT and Future HTS Systems," *Plenary Session of 2013 14th IEEE International Vacuum Electronics Conference (IVEC)*, May 21–23, 2013, Paris, France.

[8]   Article 5, *Frequency Allocations*; Chapter 2, Frequencies; Volume 1, Articles; Radio Regulations, Edition of 2016, International Telecommunication Union, Geneva, Switzerland, ISBN-10 : 9261199976, ISBN-13 : 978-9261199975.

[9]   *ECC Decision (06)03*, Electronic Communications Committee, CEPT, approved March 24, 2006, amended: March 8, 2019.

[10]   Eutelsat site, "Broadband, Extending Coverage to Bridge the Digital Divide," https:// www.eutelsat.com/en/satellite-communication-services/satellite-internet-broadband -service.html.

[11]   Fenech, H., et al., "Eutelsat HTS Systems," in "Ka Band and High Throughput Satellites," special issue of the *International Journal on Space Communications*, Vol. 34, No. 4, July/August 2016.

[12]   Eutelsat press release, "Eutelsat Selects Gilat Technology to Power Satellite Broadband Services in Western Russia," April 21, 2016.

[13]   Eutelsat press release, "TricolorTV Launches Broadband Services in Russia Via Eutelsat 36c Satellite," September 8, 2016, https://news.eutelsat.com/pressreleases/tricolortv -launches-broadband-services-in-russia-via-eutelsat-36c-satellite-1551511#: -:text=Paris%2C%208%20September%202016%20%E2%80%93%20 TricolorTV,NYSE%20Euronext%20Paris%3A%20ETL).

[14]   Hill, J., and R. Jewett, RSCC, *Hughes Officials Detail Stressful Express-AM6 Recovery Mission*, Via Satellite, May 27, 2020. https://www.satellitetoday.com/business /2014/10/22/express-am6-launch-revives-rsccs-constellation-renewal-efforts/.

[15]   Eutelsat press release, "Eutelsat 65 West A Satellite Goes Live At 65° West," May 02, 2016, https://news.eutelsat.com/pressreleases/eutelsat-65-west-a-satellite-goes-live -at-65deg-west-1389846.

[16]   Hughes website, "Hughes 65 West," https://www.hughes.com/technologies/hughes -high-throughput-satellite-constellation/eutelsat-65-west-a.

[17]   Eutelsat press release, "Stargroup Selects New EUTELSAT 65 West A Satellite for StarGo Broadband in Latin America," 9 March 2016, https://news.eutelsat.com/press- releases/stargroup-selects-new-eutelsat-65-west-a-satellite-for-stargo-broadband-in-latin -america-1337858#:-:text=Launched%20today%20by%20an%20Ariane,large%20 swathes%20of%20Latin%20America.

[18]   Eutelsat press release, "Eutelsat All-Electric Eutelsat 172b Satellite Set to Transform Connectivity Landscape in Asia-Pacific," November 21, 2017, https://news.eutelsat .com/pressreleases/eutelsat-all-electric-eutelsat-172b-satellite-set-to-transform -connectivity-landscape-in-asia-pacific-2289423.

[19]   Eutelsat press release, "EUTELSAT 172B Satellite Soars into Space," June 2, 2017, https://news.eutelsat.com/pressreleases/eutelsat-172b-satellite-soars-into-space -1997707.

[20]   Fenech, H., et al., "VHTS Systems: Requirements and Evolution," *11th European Conference on Antennas and Propagation (EuCAP)*, Davos, Switzerland, March 20–24, 2017.

[21]   Eutelsat press release, "Eutelsat Selects Ground Infrastructure Providers for its Konnect Satellite Programme," May 6, 2019, https://news.eutelsat.com/pressreleases/eutelsat -selects-ground-infrastructure-providers-for-its-konnect-satellite-programme -2868809#:~:text=Paris%2C%20Washington%20DC%2C%20May%206,deploy%20 seven%20antennas%2C%20while%20Hughes.

[22]   Eutelsat press release, "KONNECT AFRICA to Connect Schools in the Democratic Republic of Congo with High Speed Internet," March 23, 2020, https://news.eutelsat. com/pressreleases/konnect-africa-to-connect-schools-in-the-democratic-republic-of -congo-with-high-speed-internet-2984467.

[23]   Eutelsat press release, "Eutelsat Orders Konnect VHTS, A New-Generation Satellite to Deliver High-Speed Broadband Across Europe," April 5, 2018, https://news.eutelsat. com/pressreleases/eutelsat-orders-konnect-vhts-a-new-generation-satellite-to -deliver-high-speed-broadband-across-europe-2469821.

[24]   Wilcoxson, D., "Advanced Commercial Satellite Systems Technology for Protected Communications," *MILCOM 2011 Military Communications Conference*, Baltimore, MD, 2011, pp. 2280–2285, doi: 10.1109/MILCOM.2011.6127661.

[25]   Viasat press release, "ViaSat Sets Guinness World Records Title for the Highest Capacity Satellite," March 6, 2013, https://www.viasat.com/news/viasat-sets-guinness-world -records-title-for-highest-capacity-satellite#:~:text=Carlsbad%2C%20Calif.,October %202011%20by%20Viasat%20Inc.

[26]   Viasat press release, "ViaSat Introduces SurfBeam 2 Broadband Networking System," July 11, 2011, https://www.viasat.com/news/viasat-introduces-surfbeam-2-broadband -networking-system.

[27]   Henry, C., "Viasat Preps Big Insurance Claim for ViaSat-2 Antenna Anomaly," *Space News*, May 30, 2018, https://spacenews.com/viasat-preps-big-insurance-claim-for -viasat-2-antenna-anomaly/.

[28]   Viasat, Attachment A, Technical Information to Supplement Schedule S, *Application for space station authorization for ViaSat 2 to the FCC.*

[29]   Price, B., "Evolution and Future of Broadband Satellite Services," *ITU Satellite Symposium*, organized by the ITU, Geneva, Switzerland, November 29, 2018

[30]   Gunter website, Jupiter 1/EchoStar 17, https://space.skyrocket.de/doc_sdat/jupiter-1 .htm.

[31]   Alchetron website, EchoStar XVII, https://alchetron.com/EchoStar-XVII#echostar -xvii-e38a5160-b132-4b10-8b96-90c9f1ecabc-resize-750.png.

[32]   Hughes press release, *Hughes Announces HughesNet Gen5 High-Speed Satellite Internet Service*, March 7, 2017

[33]   Hughes website, Hughes Announces HughesNet Gen5 High-Speed Satellite Internet Service, https://www.hughes.com/technologies/hughes-high-throughput-satellite -constellation/echostar-xix.

[34]   Hughes, Attachment, Technical Information to Supplement Schedule S, *Application for space authorization for Jupiter 97W (Jupiter 2) to the FCC*.

[35]   Hughes press release, *EchoStar XIX Satellite with JUPITER High-Throughput Technology Successfully Positioned in Orbital Slot*, January 10, 2016, https://www.hughes.com /resources/press-releases/echostar-xix-satellite-jupiter-high-throughput-technology -successfully#:~:text=%E2%80%94Hughes%20Network%20Systems%2C%20 LLC%20(,into%20its%20permanent%20geosynchronous%20orbital.

[36]   Satellite Internet for Energy Exploration: Oil, Gas, and Mining, http://oilfieldsatellite .net/.

[37]   Hughes press release, "Hughes Launches World's Largest and Fastest Broadband Satellite Network," March 7, 2017, https://www.hughes.com/resources/press-releases/hughes -launches-worlds-largest-and-fastest-broadband-satellite-network#:~:text=Hughes%20 Launches%20World's%20Largest%20and%20Fastest%20Broadband%20Satellite%20 Network,-March%2007%2C%202017&text=Washington%2C%20 D.C.%E2%80%94Hughes%20Network%20Systems,and%20fastest%20satellite%20 broadband%20network.

[38]   Inmarsat website, *Our Satellites*, https://www.inmarsat.com/about-us/our-technology /our-satellites/.

[39]   Inmarsat press release, "Successful Launch Confirmed for Inmarsat's First Global Xpress® Satellite (Inmarsat-5 F1)," December 9, 2013, https://www.inmarsat.com/news /successful-launch-first-global-xpress-satellite/.

[40]   Inmarsat press release, "Successful Launch Confirmed for Inmarsat's Second Global Xpress Satellite (Inmarsat-5 F2)," February 2, 2015, https://www.inmarsat.com/press-release/successful-launch-confirmed-inmarsats-second-global-xpress-satellite-inmarsat -5-f2/.

[41]   Inmarsat press release, "Inmarsat Confirms Successful Launch of the Third Global Xpress (GX) Satellite," August 29, 2015, https://www.inmarsat.com/news/inmarsat -confirms-successful-launch-of-the-third-global-xpress-gx-satellite/#:~:text=29%20 August%202015%3A%20Inmarsat%20has,Global%20Xpress%20(GX)%20 constellation.

[42]   Inmarsat press release, "Inmarsat Confirms Successful Launch of the Fourth Global Xpress Satellite," May 16, 2017, https://www.inmarsat.com/news/inmarsat-confirms -successful-launch-fourth-global-xpress-satellite/.

[43]   Arianespace Launch Kit, VA250, TIBA-1 and Inmarsat GX5, November 2019.

[44]   Thales Alenia Space press release, "Thales Alenia Space Signed a Contract with Inmarsat for the Construction of GX Satellite," June 6, 2017, https://www.thalesgroup.com/en /worldwide/space/press-release/thales-alenia-space-signed-contract-inmarsat -construction-gx-satellite.

[45]   Airbus Defence and Space Press Release, "Airbus Defence and Space Signs Contract with Inmarsat to Build Two Next Generation Mobile Communications Satellites," December 24, 2015, https://www.airbus.com/newsroom/press-releases/en/2015/12 /airbus-defence-and-space-signs-contract-with-inmarsat-to-build-two-next-generation -mobile-communications-satellites.html.

[46]   Airbus Defence and Space press release, "Airbus Wins Three Satellite Deal from Inmarsat for Revolutionary Spacecraft," May 30, 2019, https://www.airbus.com/newsroom /press-releases/en/2019/05/airbus-wins-three-satellite-deal-from-inmarsat-for -revolutionary-spacecraft.html.

[47]   Inmarsat, "Inmarsat plc Reports Interim Results 2019, A Robust Performance," https:// www.inmarsat.com/wp-content/uploads/2019/12/Inmarsat_H1_2019_Results_RNS .pdf.

[48]   Arianespace Launch Kit, *VA226, Sky Muster and ARSAT-2*, September 2015.

[49]   Arianespace Launch Kit, VA231, Sky Muster II and GSAT-18, October 2016.

[50]   Ovum, "Satellite Broadband: A Global Comparison," a report for NBN™, April 28, 2016.

[51]   NBN Co website, "Sky Muster™ 101: Everything You Need to Know," July 25, 2016, https://www.nbnco.com.au/blog/the-nbn-project/nbn-sky-muster-satellite-101.

[52]   Viasat press release, "NBN Co Selects Viasat to Provide Satellite Internet System in Australia," July 12, 2012, https://www.viasat.com/news/nbn-co-selects-viasat-provide -satellite-internet-system-australia.

[53]   Airbus Defence and Space press release, "Airbus Defence and Space to Build the SES-12 Satellite," July 7, 2014, https://www.airbus.com/newsroom/press-releases/en/2014/07 /airbus-defence-and-space-to-build-the-ses-12-satellite.html.

[54]   SES press release, "SES-12: Elevating Experiences Today," May 25, 2018, https://www .ses.com/newsroom/ses-12-elevating-experiences-today.

[55]   SES press release, "SES-12 Goes Operational to Serve Asia-Pacific and the Middle East," 26 Feb 2019, https://www.ses.com/press-release/ses-12-goes-operational-serve -asia-pacific-and-middle-east#:~:text=1%20min%20read-,SES%2D12%20Goes%20 Operational%20to%20Serve%20Asia,Pacific%20and%20the%20Middle%20 East&text=Luxembourg%2C%2026%20February%202019%20%E2%80%93%20 SES,Pacific%20and%20the%20Middle%20East.

[56]   Arianespace launch kit, VA241, SES 12 and Al Yah 3, January 2018.

[57]   SES press release, "SES-14 Goes Operational to Serve the Americas," 4 September 2018, https://www.ses.com/ses-14-goes-operational-serve-americas-0.

[58]   Arianespace launch kit, VS17, SES 15, May 2017.

[59] SES press release, "SES-15 Enters Commercial Service to Serve the Americas," 15 January 2018, https://www.ses.com/press-release/ses-15-enters-commercial-service -serve-americas#:~:text=Luxembourg%2C%2015%20January%202018%20 %E2%80%93%20SES,orbital%20position%20since%201%20January.

[60] Arianespace press release, "SES Selects Arianespace for Launch of SES-17," 12 September 2017, https://www.arianespace.com/press-release/ses-selects-arianespace-for-launch-of -ses-17/#:~:text=SES%20has%20selected%20Arianespace%20to,and%20 Arianespace%20in%20Paris%20today.

[61] SES press release, "SES Orders High Throughput Satellite from Thales with First Secured Anchor Customer for Inflight Connectivity," 12 September 2016, https://www .ses.com/press-release/ses-orders-high-throughput-satellite-thales-first-secured-anchor-customer-inflight#:~:text=2%20min%20read-,SES%20ORDERS%20HIGH%20 THROUGHPUT%20SATELLITE%20FROM%20THALES%20WITH, ANCHOR%20CUSTOMER%20FOR%20INFLIGHT%20CONNECTIVITY &text=First%20long%2Dterm%20anchor%20agreement,service%20provider% 20in%20the%20Americas.

[62] Arianespace Launch Kit, VA228, Intelsat 29e, January 2017.

[63] Arianespace Launch Kit, VA235, Sky Brazil-1 and Telkom 3S, February 2017.

[64] Arianespace Launch Kit, VA232, Intelsat 33e and Intelsat 36, August 2016.

[65] Graham, W., "SpaceX Falcon 9 Launches with Intelsat 35e at the Third Attempt," NASA SpaceFlight.com July 3, 2017, https://www.nasaspaceflight.com/2017/07 /spacex-launch-surge-falcon-9-launch-intelsat-35e/.

[66] Arianespace Launch Kit, VA239, Intelsat 37e and BSAT-4a, September 2017.

[67] Arianespace Launch Kit, VA243, Horizon 3e and Azerspace-2/Intelsat 38, September 2018.

[68] Intelsat press release, "Intelsat 33e, the Second Intelsat Epic[NG] Satellite, Successfully Launched into Orbit," August 24, 2016, http://www.intelsat.com/news/press-release /intelsat-33e-the-second-intelsat-epicng-satellite-successfully-launched-into-orbit/.

[69] Intelsat press release, "Intelsat 35e Launches Successfully," July 6, 2017, http://www .intelsat.com/news/press-release/intelsat-35e-launches-successfully/.

[70] Intelsat press release, "Intelsat 37e Preparing for Launch," August 30, 2017, http://www .intelsat.com/news/press-release/intelsat-37e-preparing-for-launch/.

[71] Intelsat website, "Horizons 3e at 169°E," http://www.intelsat.com/global-network/ satellites/fleet/horizons-3e/.

[72] Intelsat press release, "Maxar Technologies Will Build Next-Generation Intelsat Epic Geostationary Communications Satellite with NASA Hosted Payload," February 3, 2020, https://www.intelsat.com/newsroom/maxar-technologies-will-build-next -generation-intelsat-epic-geostationary-communications-satellite-with-nasa-hosted-payload/.

8

Non-Geostationary Systems

8.1 Introduction

Chapters 1–7 focus on GEO HTS systems, but they are not the only solution for delivering data and Internet services via satellites. Accordingly, this chapter looks at non-GEO systems, which have received considerable interest in recent years. Communication LEO systems have been around since the 1990s but have seen increased activity in the last decade.

Section 8.2 examines the different orbit solutions that are available to construct a constellation, reviewing the physics that govern them and the limitations of the options. Section 8.3 investigates the HTS link budget to bring out the salient aspects of the LEO case in comparison with the MEO and GEO cases. Next, Section 8.4 provides an outline of the considerations in the designing and dimensioning of a constellation. Finally, Section 8.5 discusses existing systems, systems that have been proposed, and proposed systems that are still in the making.

8.2 Orbits

Kepler postulated that all orbits involving two bodies are elliptical, with the heavier body being at the focus. This means that the orbital plane of the satellite must pass through the center of the heavier body. Circular orbits are thus a specific case. Planets rotating around the Sun and artificial satellites revolving around the Earth are governed by the same orbital mechanics.

The third law of Kepler defines the sidereal period of an orbit, which is defined as

$$T = 2\pi \sqrt{\frac{a^3}{\mu}}$$ (8.1)

where a is the semi-major axis in meters, and μ is the standard gravitational parameter for Earth and equal to 3.986×10^{14} m^{3}s^{-2}.

When we consider a satellite, it is convenient to rewrite (8.1) as

$$T = 2\pi \sqrt{\frac{\left(R_E + A_S\right)^3}{\mu}}$$ (8.2)

where R_E is the Earth's radius and A_S is the altitude of the satellite.

The period as seen by an observer on Earth, the synodic period, will either be longer if the satellite orbit is prograde (i.e., both the satellite and Earth are rotating in the same direction) or shorter if the satellite orbit is retrograde (i.e., the satellite and Earth are rotating in opposite directions).

The Earth's equatorial radius is generally taken to be 6,378 km, and this is the figure often used for a satellite in GEO. However, the Earth is flattened at the poles where the radius is 6,357 km. The average radius of the Earth is 6,371 km, and this may be a more appropriate figure when looking at inclined orbits. Figure 8.1 depicts the variation of the sidereal period of a satellite in a circular orbit as a function of the altitude; it also gives the mean orbital velocity required to maintain a stable orbit. As shown in Figure 8.1, the period increases with altitude while the orbital velocity for a stable orbit decreases with altitude.

The following four broad classes of orbits can be identified:

- *LEOs* have altitudes between 200 and 2,000 km. However, different definitions exist. NASA [1, 2] defines LEO altitudes as from the atmosphere-space interface (which NASA defines as 80 km) to 2,000 km; the Inter-Agency Space Debris Coordination Committee [3] defines them as starting from the Earth's surface to 2,000 km; and the European Space Agency (ESA) [4] defines LEOs as having altitudes normally below 1,000 km and down to as low as 160 km.

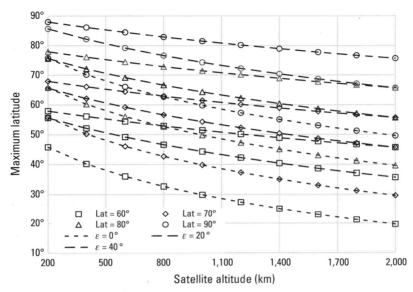

Figure 8.1 Plot of the sidereal period and the mean orbital velocity for a stable orbit of a satellite in a circular orbit at an altitude above Earth.

The lower operational limit is constrained by the outer layers of the atmosphere, the thermosphere, and the exosphere, depending on the altitude. The atmosphere exhibits periodic variations over latitudes. Although the density of the atmosphere is very low at these altitudes, there is still aerodynamic drag that would cause orbit decay; therefore, satellite propulsion resources are necessary to maintain altitude. High LEO may overlap with the inner Van Allen belt, which includes both energetic electrons and protons [57]. On the equatorial plane, the maximum omnidirectional flux is 100 $cm^{-2}s^{-1}$ for protons with an energy of more than 10 MeV at an altitude of about 1,000 km, but this increases by four orders of magnitude in 2,000 km thereafter. Over the same altitude range, the flux for electrons with more than 1 MeV of energy is approximately one order of magnitude less. The inner Van Allen belt is roughly toroidal and slightly concave toward the Earth so that the fluence decreases rapidly for protons above 45° to the equatorial plane. The International Space Station is in an inclined LEO, and most Earth observation satellites are in a Sun-synchronized polar LEO. Communication constellations are in inclined LEO. Thus, the integrated radiation dosage varies accordingly.

The Earth's magnetic field is weaker in the South America and the South Atlantic regions. This causes the inner Van Allen belt to get closer to Earth in these regions, producing a lower altitude area with a high flux of charged particles. This is called the South Atlantic Anomaly. This phenomenon has been identified as a possible cause for the anomalies on Globalstar satellites.

Since such orbiting satellites perform an orbit in a short period relative to a sidereal day, the visibility of each satellite by a fixed terrestrial observer is limited to a few minutes; several satellites per plane and several planes are required for continuous operation, but the resulting coverage covers all longitudes. These systems lend themselves to global application. Typically, the constellations include hundreds or thousands of small satellites.

- *Medium-Earth orbits (MEOs)* are often defined as orbits with altitudes above 2,000 km and below GEO. This range of altitudes includes the regions with the highest intensity of particles for both the inner and the other Van Allen radiation belts. The inner belt peaks at about 4,500 km at the equator with a maximum omnidirectional flux of 3.4×10^7 cm^{-2}s^{-1} for 10 MeV protons, while the outer belt peaks with a flux of 4×10^6 cm^{-2}s^{-1} for 1 MeV for electrons at about 21,700 km [5]. The outer belt can be approximated as toroidal, so the fluxes decrease as the orbit inclination is increased. Most of the satellite activity concentrates at the semisynchronous orbit with a period of 12 h. NASA [2] defines the semisynchronous orbit as 20,200 km, ±300 km. Most global navigation satellite systems operate around the semisynchronous orbit. MEOs share the constellation concepts of LEOs, since several planes with several satellite per plane are required, although the satellites are at higher altitudes, which also helps reduce the constellation size to a few tens. Examples include the global positioning system (GPS), Galileo, and Glonass with altitudes between 19,130 and 23,222 km and inclinations between 55° and 64.8°. The O3B constellation operates in an equatorial at an altitude of 8,062 km.

- *GEO* is a unique circular orbit (with zero eccentricity) in the equatorial plane (with zero inclination) with a periodicity of one sidereal day or 23 hours, 56 minutes, and 4 seconds. Thus, a GEO satellite will appear stationary to a terrestrial observer as there is no relative motion. NASA [2] defines it as having an altitude of 35,788 km, ±300 km while the Inter-Agency Space Debris Coordination Committee [3] defines the GEO region as an altitude of 35,786 km ±200 km and with an orbital inclination of ±15°. GEO lies toward the edge of the Van Allen belts where diurnal and short-term variations of orders of magnitude can occur. GEO is also subject to solar protons and cosmic rays. This orbit is home to many communication satellites capitalizing on the fact that it is geostationary and therefore requiring simple Earth terminals with no operational steering required.

• *Highly elliptical orbits (HEOs)* are orbits with a high eccentricity and usually inclined at 63.4°. The oblateness of the Earth creates a drift in the argument of the perigee or rotation with time of the apogee within the orbital plane inclination. The selected inclination produces zero drift and therefore requires minimal propellant for the correction. The ellipticity prolongs the dwell duration at the apogee and therefore can be used a quasi-stationary satellite for a duration. When the argument of the perigee is 270°, the apogee and thus the associated service area are over the northern hemisphere while when it is 90°, the apogee is over the southern hemisphere. There are two well-known orbits. The Molniya orbit has a period of half a sidereal day and is typically a constellation of three phased satellites with each satellite being used for eight hours centered on the apogee. Because the period of Molniya is half a day, it has it has two apogee dwell regions separated in longitude by 180°. Molniya orbits were used by Russia, as they allow high-elevation communications to be maintained at high latitudes (above the limitation of GEO). The high inclination also reduces the launch requirements especially when launching from high latitude, which is appropriate for the Russian launch sites. The second dwell location over the United States was also an interesting feature for the Russians! The Molniya orbit passes four times a day through the Van Allen belts. The Tundra orbit is similar but with an orbit of one sidereal day so that it has one dwell area. It requires a constellation of two satellites for continuous coverage and avoids the Van Allen belts. Russia still operates the Meridian satellites in the Molniya orbit [8] and the first Sirus satellites operated in the Tundra orbit.

8.3 The HTS Link Budget

This analysis will initially maintain the same cell dimension of the Earth's surface for different scenarios where the satellite altitude is different. Looking at Figure 8.2, this means that for whatever satellite orbit, the geocentric angle,

θ_E, corresponding to the service area dimension, is constant. Further, the cell dimensions considered make the geocentric angle small, and the approximation of sine of a small angle can be applied.

From Figure 8.2, we can relate the angle subtended from the satellite to cover the cell to the geocentric angle as

$$\theta_S = 2 \operatorname{atan}\left(\frac{R_E \theta_E}{2 A_S}\right) \tag{8.3}$$

where θ_E is the corresponding geocentric angle to cover the cell.

In Figure 8.2, the satellite altitude, A_S, is shown as A_{LEO}, A_{MEO} and A_{GEO} for the LEO, MEO, and GEO as the case may be. Also note that in Figure 8.2, the left part of the beam is not shown for clarity. This is assumed to be symmetric to the right side.

For simplicity, we assume the Earth to be a sphere, but the analysis can be refined to take into account the elevation of the observer due to the topology of the Earth or the altitude of an aircraft for that matter. The following transformations apply [9]:

$$\begin{aligned} R_E &\to R_E + A_u \\ A_S &\to A_S - A_u \\ R_E + A_S &\to R_E + A_S \end{aligned} \tag{8.4}$$

where A_u is the elevation or altitude of the user with respect to the mean sea level.

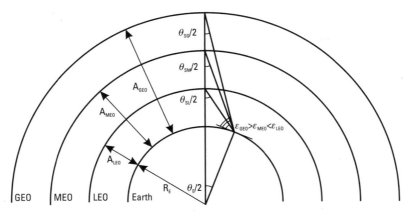

Figure 8.2 Diagram showing the same cell covered from satellites in LEO, MEO, and GEO (not to scale).

Note that R_E is the mean sea-level Earth radius as described with respect to the reference geoid so that $R_E + A_U$ gives the distance from the user to the center of the Earth, $A_S - A_U$ gives the distance from the user to the satellite, and $R_E + A_S$ gives the distance from the satellite to the center of the Earth.

Thus, for a given cell, the required antenna aperture diameter to cover the cell can be expressed as

$$D_{ANT} = \frac{Kk_{ANT}\lambda}{2\operatorname{atan}\left(\dfrac{R_E\theta_E}{2A_S}\right)} \tag{8.5}$$

where k_{ant} is an antenna constant (as discussed in Chapter 5), and λ is the wavelength of the considered frequency.

The corresponding antenna gain for a circular aperture can be expressed as

$$
\begin{aligned}
G_{ANT} &= \eta\left(\frac{\pi D_{ant}}{\lambda}\right)^2 \\
&= \eta\left(\frac{\pi k_{ant}}{2\operatorname{atan}\left(\dfrac{R_E\theta_E}{2A_S}\right)}\right)^2
\end{aligned}
\tag{8.6}
$$

The *C/N* at the terminal at the satellite is given by the link budget and often expressed in decibels as

$$\left(\frac{C}{N_0}\right)_{dB} = -(kT)_{dB} + P_{HPA,Sat,dB} + L_{PL,dB} + G_{ant,sat,dB} + L_{FSL,dB} + \frac{G}{T}_{ter,dB} \tag{8.7}$$

or as

$$
\begin{aligned}
\frac{C}{N_0} &= \frac{1}{kT}P_{HPA,sat}\,L_{PL}\,G_{ant,sat}L_{FSL}\frac{G}{T}_{ter} \\
&= \frac{1}{kT}P_{HPA,sat}\,L_{PL}\,\eta\left(\frac{\pi k_{ant}}{2\operatorname{atan}\left(\dfrac{R_E\theta_E}{2A_S}\right)}\right)^2\left(\frac{\lambda}{4\pi R_S}\right)^2\frac{G}{T}_{ter}
\end{aligned}
\tag{8.8}
$$

where:

k is a Boltzmann constant and equal to -228.6 dB W/K/Hz;

T is the system temperature in Kelvin;

$P_{\text{HPA,sat}}L_{\text{pl}}G_{\text{ANT,sat}}$ is the satellite EIRP with $P_{\text{HPA,sat}}$ being the payload available RF power, L_{PL} the output losses, and $L_{\text{PL}} < 1$ and $G_{\text{ANT,sat}}$ the satellite peak antenna gain;

L_{FSL} is the free-space loss between the satellite and the terminal and $L_{\text{FSL}} < 1$;

G/T_{Ter} is the terminal G/T.

The decibels in the subscript indicate the value expressed in decibels; otherwise the symbol represents the absolute value. (Note that the losses in decibels are negative values.)

Further, we consider all parameters to be fixed except the satellite antenna gain and the free-space loss so as to fix the cell dimension on Earth as the altitude changes. Thus, lumping all constant parameters in K_{LB}, we get

$$\frac{C}{N_0} = K_{\text{LB}}\left(\frac{1}{\text{atan}\left(\dfrac{R_E\theta_E}{2A_S}\right)}\right)^2\left(\frac{1}{R_S}\right)^2 \tag{8.9}$$

When the argument of the arctangent is small, we can use the approximation of a tangent of a small angle and at the satellite nadir slant range becomes equal to the satellite altitude. Thus, (8.9) simplifies further to

$$\frac{C}{N_0} = K_{\text{LB}}\left(\frac{1}{\left(\dfrac{R_E\theta_E}{2}\right)}\right)^2 \tag{8.10}$$

Equation (8.10) shows that at the subsatellite point when the cell size, θ_E, being constant, the link budget is independent of the satellite altitude. The beamwidth of the satellite antenna is a function of the cell size, θ_E, and the altitude A_S. The approximation for small arguments of a tangent implies that $R_E\theta_E \ll 2A_S$, which is a function of both the cell size and the altitude, so care is required at low altitudes and/or large cell dimensions. Thus, if assume θ_E to be such that the corresponding angle from GEO, θ_{SG}, is 0.5°, for satellite altitudes above 400 km and 550 km, the error is less than 10% and 5%, respectively.

Deviating from the subsatellite point to the edge of the service area, the slant range has to be considered and can be expressed invoking the law of cosines as applied to Figure 8.1 as

$$R_S = \sqrt{R_E^2 + \left(R_E + A_S\right)^2 - 2R_E\left(R_E + A_S\right)\mathrm{Cos}\left(\frac{\theta_S}{2}\right)} \qquad (8.11)$$

It is possible to plug (8.11) into (8.9) for the slant range. Depending on the satellite system, the antenna gain needs to be updated accordingly. If the edge of the service area corresponds to the eoc of the beam, then the delta gain of the antenna needs to be considered, but if the beam is steered in that direction so that the beam boresight is positioned at the edge of the service area, no modification is required since the edge of the service area corresponds to the beam center.

We consider three HTS link budget analysis scenarios: GEO, MEO, and LEO. HEO is not considered further in this section since it is not ideal for HTS applications. The reference case is a GEO system, and the chosen case is based on the case studied in Chapter 5. We assume that the satellite nadir corresponds to the beam boresight. For each scenario, we look at the forward link, and we limit ourselves to the downlink, since it is the most dimensioning from the satellite point of view. Table 8.1 provides this analysis.

The aim of the analysis is to assess the satellite resources that are required to maintain the same C/N_0 so that each scenario provides the same signal quality for a given terminal in the same service area. The operational frequency in all cases is assumed to be the same.

Recall that the first step is to maintain the same cell size. The geocentric angle of the beam, θ_E, is thus constant. This is illustrated in Figure 8.2.

The first three columns of Table 8.1(a) assume a given cell of about 300 km. This is equated to the 4-dB contour of the antenna and a beamwidth of 0.46° at GEO as considered in Chapter 5. The available TWTA RF power is assumed to be constant in all three scenarios and equal to 100-W RF. The antenna efficiency and the payload output losses are assumed to be identical in all cases. As the altitude decreases, the beamwidth of the antenna required to cover a given cell increases so that the antenna gain decreases. To meet these requirements, the antenna dimensions get smaller as the altitude is decreased.

The antenna aperture diameter that for GEO was 2.6m scales down to 0.04m at LEO. This is encouraging since LEO satellites tend to be smaller with, for example, a square meter available for the antenna accommodation.

Table 8.1
A Simplified Satellite EIRP Budget and the Downlink Forward Budget for Three
Scenarios: GEO, MEO, and LEO

Satellite	GEO	MEO	LEO	
Antenna diameter	2.6m	1.4m	0.04m	0.4m
Frequency		19.70 GHz		
Beamwidth (4 dB)	0.46°	0.84°	28.6°	3.0°
Cell diameter	297 km	297 km	297 km	31 km
Antenna efficiency		65%		
Antenna gain	52.7 dBi	47.4 dBi	15.2 dBi	36.5 dBi
HPA power	100.0W	100.0W	100.0W	1.1W
Output losses		1.7 dB		
EIRP (peak)	71.0 dBW	65.7 dBW	35.2 dBW	35.2 dBW
EIRP (eoc)	67.0 dBW	61.7 dBW	31.2 dBW	31.2 dBW

(a)

Link				
Satellite EIRP	71.0 dBW	65.7 dBW	35.2 dBW	35.2 dBW
Range	37,236 km	20,200 km	600 km	600 km
Free-space loss	209.8 dB	204.4 dB	173.9 dB	173.9 dB
Earth station G/T	16.1 dB/K	16.1 dB/K	16.1 dB/K	16.1 dB/K
C/N_0	106.0 dBHz	106.0 dBHz	106.0 dBHz	106.0 dBHz

(b)

If we now optimize the usage of the available accommodation area on a LEO platform, we can afford to increase the antenna aperture and decrease the cell dimension. Additionally, it is also possible to lower HPA power. This is shown in the last column of Table 8.1, where with a 40-cm aperture, the cell size is reduced to about 30 km and requires about a 1-W RF. Of course, there is no magic since if we consider the original 300-km cell as the service area for the scenario in column four and populate the service area with the new cells with a dimension of 30 km, there would be about 91 such cell with a total RF power requirement of 100W!

Smaller cells mean that there is more scope for the management of the satellite resources, since in HTS, the downlink information is, in general, intended for a single user so that larger cells may require a simpler antenna solution but are inefficient power-wise. For example, not all cells may be equally busy at any given time so that the available power may be apportioned according to the cell requirement at any given time. Of course, this management layer also requires satellite power. For example, routing and beamforming require power, which becomes another entry in the power budget, so the balance between intelligence power and RF power becomes an important consideration.

Figure 8.2 shows the elevation, ε, as ε_{LEO}, ε_{MEO}, and ε_{GEO} for the LEO, MEO, and GEO cases, respectively, and demonstrates that if the cell dimension is constant the elevation angle is not. From Figure 8.3, we can apply the law of sines and write

$$\frac{Sin\left(\dfrac{\pi}{2}+\varepsilon\right)}{R_E+A_S}=\frac{Sin(\beta)}{R_E} \tag{8.12}$$

where ε is the elevation angle from the terminal to the satellite.

Also, from Figure 8.3, we can express

$$\beta=\pi-\frac{\pi}{2}-\varepsilon-\frac{\theta_E}{2} \tag{8.13}$$

Using (8.12) and (8.13) together with the expansion $Cos(A+B)$ we get [9]

$$\varepsilon=Atan\left(Cot\left(\frac{\theta_E}{2}\right)-\frac{R_E}{\left(R_E+A_S\right)Sin\left(\dfrac{\theta_E}{2}\right)}\right) \tag{8.14}$$

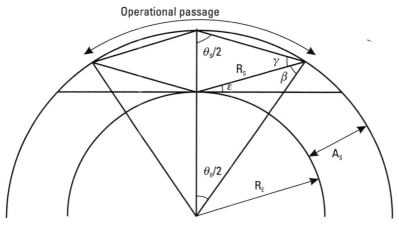

Figure 8.3 Geometry of the operational passage of a satellite.

The elevation in a GEO is not meaningful in this context, as theoretically, it is not a function of the satellite ephemeris but of the relative position of the observer and the satellite. Thus, at the subsatellite point the elevation is 90° while for a circle on Earth passing through 81°N and 81°S latitude at the satellite longitude, the elevation is 0°. This elevation is maintained for a circle on Earth with these two points as diameter and the subsatellite point as the center.

The minimum elevation of the satellite from an observer on Earth over the passage of a LEO satellite is a function of the cell dimension. The geometry for LEO is similar to that of GEO but dynamic. As can be seen from Figure 8.3, for a static observer on the orbital plane, the satellite first appears over the horizon (at 0° elevation) but at this stage it is probably not within the service area. The elevation increases so that according to the defined service area, the minimum operational elevation is reached when the service area reaches the observer. The elevation then increases to 90° when the satellite passes over the observer, and then the sequence repeats itself in the reverse order.

Of course, the observer does not happen to be on the orbital plane for each satellite passage. A constant elevation is a circle with the satellite subsatellite point as a circle, and by definition the dimensions are dictated by the service area, which determines the minimum operational elevation. Thus, when the satellite service area passes over the observer, the elevation starts at the minimum operational elevation and increases to maximum before decreasing again. The maximum elevation is a function of the orthogonal distance of the observers on Earth from the orbital plane.

The minimum operational elevation is given in Figure 8.4 as a function of the satellite altitude with the service area diameter as a parameter. One can deduce that as the service area increases, the elevations decrease since there is a downward shift in the plot with increasing service area dimensions. Additionally, the elevations decrease with decreasing satellite altitudes. In general, there is a positive gradient with satellite altitude, and the variation of elevations becomes more significant for lower satellite altitudes. For comparison, the range of elevation for the equivalent service area dimensions for semisynchronous MEO is 78.2° to 88.8°.

Referring to Figure 8.3 and using the law of cosines, we can write the following expression, which relates the slant range, the elevation, and the satellite altitude:

$$\left(R_E + A_S\right)^2 = R_S^2 + R_E^2 - 2R_S R_E \cos\left(\frac{\pi}{2} + \varepsilon\right) \tag{8.15}$$

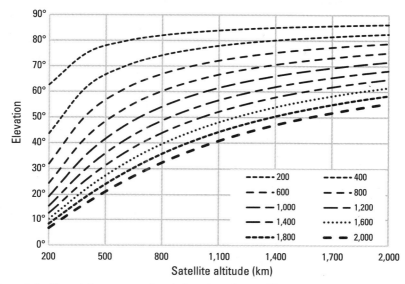

Figure 8.4 The minimum operational elevation of a satellite at a given altitude for an observer on Earth with the service area dimension as the parameter.

As (8.15) is a quadratic equation of the slant range, mathematically the slant range can be expressed as

$$R_S = -R_E \operatorname{Sin}\varepsilon \pm \sqrt{\left(R_E \operatorname{Sin}\varepsilon\right)^2 + A_S^2 + 2R_E A_S} \qquad (8.16)$$

Since the slant range has a positive value, only the positive option is valid so that (8.16) becomes.

$$R_S = -R_E \operatorname{Sin}\varepsilon + \sqrt{\left(R_E \operatorname{Sin}\varepsilon\right)^2 + A_S^2 + 2R_E A_S} \qquad (8.17)$$

The slant range is lowest when the elevation is 90° so that the variation in the free-space loss during the operational passage can be written as

$$20\operatorname{Log}\left(\frac{R_S}{A_S}\right) = \frac{-R_E}{A_S}\operatorname{Sin}\varepsilon + \sqrt{\frac{\left(R_E \operatorname{Sin}\varepsilon\right)^2}{A_S^2} + 1 + \frac{2R_E}{A_S}} \qquad (8.18)$$

Figure 8.5 depicts (8.17) and (8.18) for altitudes between 200 and 2,000 km and for elevations between 0° and 40°.

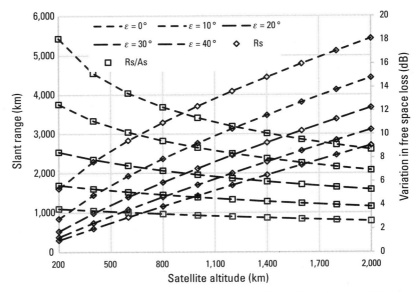

Figure 8.5 Plot giving the slant range for a given satellite altitude between 200 and 2,000 km, with the elevation as a parameter between 0° and 40°, and the variation of the free-space loss over the operational passage.

Notice that the slant range increases faster with satellite altitude for lower elevations while the variation of the free-space loss increases rapidly for decreasing altitude. This effect is practically negligible at high altitude. For example, for the semisynchronous MEO, the worst-case figure for the range of parameter considered is 2.1 dB, while at GEO the equivalent figure is 1.3 dB.

Latency is an important parameter in digital communications, and Figure 8.6 shows how the round-trip delay varies with altitude. The round-trip delay drops from about 250 ms for GEO, to 135 ms for a semisynchronous MEO, to 4 ms for a 500-km LEO.

This is an intrinsic advantage of LEO systems over other orbit options.

8.4 The Constellation

As shown in Figure 8.4 and the associated discussion, the service area dimensions of a satellite, the minimum operational elevation, and the satellite altitude become important system parameters since they affect the number of satellites in the constellation.

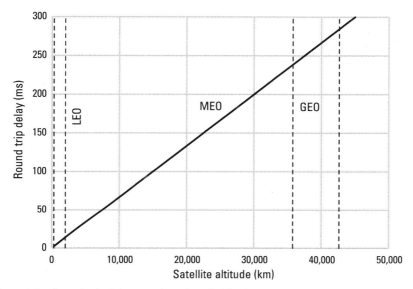

Figure 8.6   Round-trip delay as a function of altitude.

If we assume that global coverage is a requirement, the total area to be covered can be expressed as the surface area of a sphere

$$SA_E = 4\pi R_E^2 \tag{8.19}$$

The service area of each satellite has been described as a circle. Since it is not easy to pack circles in a ubiquitous fashion, we define the service area as a hexagon that can be circumscribed within the circle, similar the operation of defining cells in a GEO HTS system. Thus, the area of satellite service area can be expressed in terms of the diameter, θ_E, of the satellite service area as

$$SA_S = \frac{3\theta_E^2 R_E^2 \sqrt{3}}{8} \tag{8.20}$$

We can conclude that the total number of satellites is at least the quotient of (8.19) and (8.20), so that we get

$$N_T \geq \frac{32\pi}{3\theta_E^2 \sqrt{3}} \tag{8.21}$$

We define the operational passage of a satellite with respect to a fixed observer on Earth as the portion of the orbit where the visibility of the satellite is above the minimum operational elevation. This case, which represents the longest operational passage, is illustrated in Figure 8.3.

A typically homogenous constellation is made up of inclined orbital plane equally spaced on the equatorial plane such that the ascending nodes occur at equal intervals over $\pi/2$. (The descending nodes occupy the other $\pi/2$.) If we consider a single plane, a satellite provides service over the operational passage. Referring to Figure 8.1, rewriting (8.14), and putting the geocentric diameter of service area as the subject of the equation, we arrive at

$$\theta_E = 2\cos^{-1}\left(\frac{R_E\cos(\varepsilon)}{(R_E + A_S)} - \varepsilon\right) \tag{8.22}$$

Figure 8.7(a) illustrates the constellation concept around the equators where satellites populate planes, and several planes are used to provide a ubiquitous service area. Using the service area geometry for satellites in the same plane as given Figure 8.7(b), the phasing between two adjacent satellites in the same plane is given as

$$\Delta\varphi_S = \frac{\theta_E\sqrt{3}}{2}$$

$$= \sqrt{3}\cos^{-1}\left(\frac{R_E\cos(\varepsilon)}{(R_E + A_S)} - \varepsilon\right) \tag{8.23}$$

The relative phasing of adjacent satellites is the same for any two adjacent satellites in any given plane.

Thus, the number of satellites to fully populate the plane such that service is available 100% of the time on the ground track of that plane is given as

$$N_S = \text{Ceiling}\left(\frac{2\pi}{\sqrt{3}\cos^{-1}\left(\frac{R_E\cos(\varepsilon)}{(R_E + A_S)} - \varepsilon\right)}\right) \tag{8.24}$$

The ceiling function, which is defined as the smallest integer that is not smaller than the argument of the function, ensures that there no gaps in coverage within a plane.

Recall that several planes are required to provide continuous coverage over a given range of latitude. Because the orbits are geocentric and circular, there is symmetry between the coverages of the northern hemisphere and the

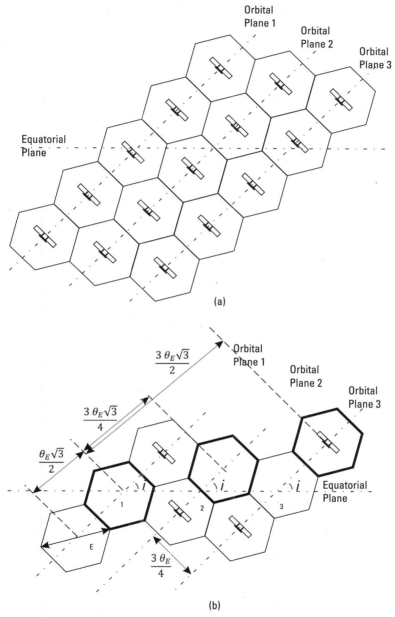

Figure 8.7 Geometry of a satellite constellation: (a) the general picture and (b) a detail showing the ascending nodes of the orbital planes and delta true anomaly for satellites in the same plane and different planes. The bold hexagon indicates the reference satellite of the plane.

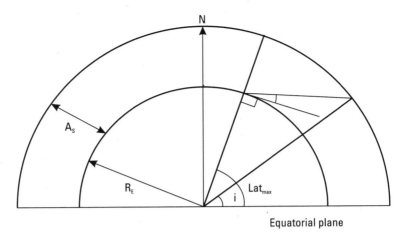

Figure 8.8 The geometry for estimating the required orbit inclination as a function of satellite altitude for a given maximum latitude coverage and a given terminal elevation.

southern hemisphere. The orbit inclination is used to control the coverages; higher orbit inclination ensures higher latitude coverage.

An estimate of the orbit inclination can be obtained as a function of satellite altitude with the maximum latitude to be covered and the required operational terminal elevation. It should be noted that the previous estimates provided the minimum terminal elevation where most of the time the operational elevation was higher. Theoretically in the case of the maximum latitude, at the limit of latitude coverage the elevation is equal to the estimate. Figure 8.8 demonstrates the associated geometry.

Applying the law of sines to Figure 8.8, we can write

$$\frac{\sin\left(\frac{\pi}{2} + \varepsilon\right)}{R_E + A_S} = \frac{\sin\left(\pi - \left[\frac{\pi}{2} + \varepsilon\right] - \left[\text{Lat}_{max} - i\right]\right)}{R_E} \tag{8.25}$$

Putting the orbit inclination as subject to the equation, we obtain

$$i = \text{Lat}_{max} + \varepsilon - \cos^{-1}\left(\frac{R_E \cos(\varepsilon)}{R_E + A_S}\right) \tag{8.26}$$

A parametric plot is provided for (8.26) in Figure 8.9 giving the orbit inclination as a function of the satellite altitude with the parameters being the maximum latitude to be covered and the terminal elevation.

Notice that there is a linear vertical shift for plots with the same terminal elevation for a given altitude and the gradients decrease with increasing terminal elevations.

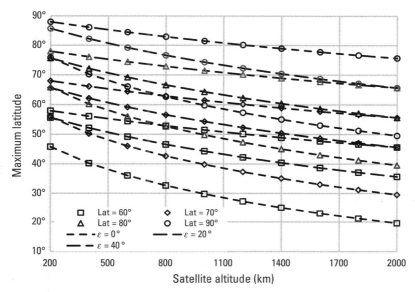

Figure 8.9 A plot of (8.26) giving the orbit inclination as a function of the satellite altitude with the maximum latitude to be covered and the terminal elevation as parameters.

To provide a ubiquitous service area around the equator, the planes need to be adjacent in a seamless fashion as shown Figure 8.7(a). The orthogonal separation between two consecutive planes at the equator as shown in Figure 8.7(b) can be expressed as

$$d_p = \frac{3\theta_E}{4} \tag{8.27}$$

Thus, the delta of ascending nodes of two consecutive planes is given as

$$\Delta\Omega = \Omega_p - \Omega_{p-1}$$
$$= \frac{3\theta_E}{4} \csc(i) \tag{8.28}$$

where Ω_p and Ω_{p-1} are the ascending nodes of the pth and $p - 1^{th}$ planes.

To ensure that the service area is ubiquitous, the number of planes can be written as

$$N_p = \text{ceiling}\left(\frac{4\pi}{3\theta_E}\right)\sin(i) \tag{8.29}$$

where i is the inclination of the orbit so that (8.28) becomes

$$\Delta\Omega = \frac{\pi}{\text{ceiling}\left(\dfrac{4\pi}{3\theta_E}\right)\sin(i)} \tag{8.30}$$

The ceiling function in (8.27) and (8.30) is used to ensure ubiquity by introducing a small overlap. However, the system architect can tweak other parameters like θ_E to minimize the overlap and enhance the capacity.

The total number of satellites in the constellation is, thus, given by (8.24) and (8.29) as

$$N_T = N_S N_P \tag{8.31}$$

Besides setting the ascending nodes, the phasing of the satellites is the adjacent plane need to set. Looking at Figure 8.7 (b), we adopt a reference satellite in each plane and define the delta phasing between the reference satellite in adjacent planes. Using the employed service area lattice, we get the relative phasing of two reference satellites in two adjacent planes as

$$\Delta\varphi_P = \frac{3\theta_E\sqrt{3}}{4} \tag{8.32}$$

8.5 Current Constellations

There are a few operational LEO communication constellations. Table 8.2 summarizes the main features of these systems.

These systems date, which back to the 1980s and 1990s, are good examples of communication constellations, although they were not conceived to deliver HTS services. All three operators went into Chapter 11 bankruptcy between 1999 and 2002 but after restructuring continue to be operational today. They deliver services like IoT, voice, and low-rate data, which require a few kilobits per second.

HTSs are broadband-oriented and intended for IP services requiring tens of megabits per second, some four orders of magnitude higher than the rates delivered by these earlier constellations. The first and only operational HTS constellation system today is O3B, which originally stood for the *other three billion* people who at the time had limited or no internet access. The system, operated by SES, is slightly specific in the sense that it is a MEO

Table 8.2
The Salient Features of Globalstar, Iridium, and Orbcomm

Name	Satellite Altitude (Km)	Number of Sat./ In-Orbit Spares/ Ground Spares	Planes	Orbit Inc.	Frequency Band	Service Area	Services
Globalstar [10]	1414	48/4/0	8	52°	S-band	70°N to 70°S	Data up to 9.6 kbps, personal com., IoT
Iridium [11]	780	66/9/6	11	86.4°	L-band	Global	Data up to 1,408 kbps, personal com., IoT
Orbcomm [12, 13]	815	32/0/0	4	45°	VHF	65°S to 65°N	Data 2.4 kbps up/ 4.8 kbps down, Machine-to-machine (M2M), IoT, automatic identification system (AIS)
	740	4/0/0	1	70°			
	975	4/0/0	1	0°			
	785–875	4/0/0	1	108°			

system in the equatorial plane (with zero nominal inclination). This implies that the coverage of the system is limited in latitude, and service is delivered in the equatorial belt between latitudes of 45°S and 45°N. The first-generation constellation included 12 satellites designed and constructed by Thales Alenia Space utilizing the ELiTeBus platform that were deployed in a circular orbit of 8,062-km altitude with an orbit period of 288 min. Eight additional satellites were subsequently ordered. The system has a latency of 150 ms. The dry mass of each satellite is 700 kg, and the satellites have a design lifetime of 10 years [14]. The satellites in production are shown in Figure 8.10. The system operates in the Ka-band: 17.8–18.6 GHz, 18.8–19.3 GHz, and 19.7–20.2 GHz for the downlink and 27.6–28.4 GHz, 28.6–29.1 GHz, and 29.5–30.0 GHz for the uplink. It is claimed to be the world's only commercially successful non-geostationary (communication) satellite system. There were five launches in groups of four, as listed in Table 8.3, to build up a fleet of 20 satellites.

Each satellite includes 12 mechanically steerable beams: 10 user beams and two gateway beams. This is evident in Figure 8.10. Each user beam is associated with a forward and return channel, each with a bandwidth of 216 MHz. Thus, each satellite has a total capacity of 4.32 GHz. It is claimed that

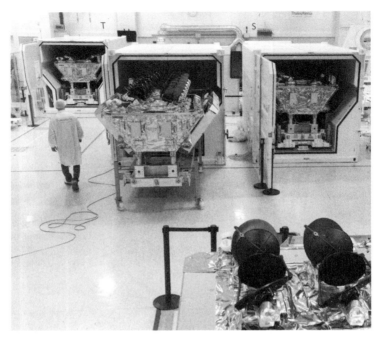

Figure 8.10   O3B satellites in production. (*Source:* Thales Alenia Space. Reprinted with permission.)

each beam has a capacity of 600 Mbps in each direction with the underlying assumption of 2.8 bps/Hz. This is, of course, very dependent on the terminals employed.

In 2017, SES [20] announced the second generation of seven satellites to be manufactured by Boeing Satellite Systems using the 702X platform with launches staring in 2021. The procurement of four additional satellites has

Table 8.3
First-Generation O3B Launches

Satellites	Launcher	Launch Date
O3B PFM, FM2, FM3, FM4 [15]	Soyuz VS05	June 25, 2013
O3B FM5, FM6, FM7, FM8 [16]	Soyuz VS08	July 10, 2014
O3B FM9, FM10, FM11, FM12 [17]	Soyuz VS10	December 18, 2014
O3B FM13, FM14, FM15, FM16 [18]	Soyuz VS18	March 9, 2018
O3B FM17, FM18, FM19, FM20 [19]	Soyuz VS05	April 4, 2019

recently been announced [21]. The satellites exploit digital payload technology to produce 5,000 electronically steerable user beams [22]. This provides considerable flexibility is allocation capacity where and when required to meet the dynamic demand for capacity. The satellite capacity is an order of magnitude higher than that of the first generation so the new generation fleet will deliver more than a 1 Tbps of capacity.

Teledesic [23, 24] was one of the first proposed (in 1994) large constellations with 840 satellites proposed in the original constellation operating the Ka-band. The system mission was global Internet access. This was reduced to 288 in 1997 with 12 planes each with 24 satellites at an altitude of 1,315 km with an inclination of 98.2°. However, on 1 October 2002, it suspended satellite production.

OneWeb was another proposed large constellation with 648 satellites orbiting at 1,200-km altitude with an orbit period of 109 min. OneWeb was producing its own satellites. Each satellite covers an area approximately the size of Alaska. The satellite lifetime is five to seven years. The OneWeb constellation operates in the Ku-band and does not include intersatellite links. OneWeb proposed the progressive pitch technique to reduce the interference to GEO satellites. The satellites are gradually tilted as they approach the equator, so that the beams are never aligned with a Ku-band GEO satellite.

By March 21, 2020, OneWeb launched 74 satellites in three batches of six, 34, and 34 [25–27], but on March 27, 2020 OneWeb filed for Chapter 11 [28]. On 27 May 2020 OneWeb [29] submitted a modification request to the U.S. Federal Communications Commission to increase the number of satellites in its constellation up to 48,000 satellite. There has been interest from a consortium including Her Majesty's Government through the U.K. Secretary of State for Business, Energy and Industrial Strategy and Bharti Global Limited [30]. Hughes has also joined the consortium [31].

The Telesat LEO constellation, proposed by Telesat, will include 298 satellites. The system capacity will be 16–26 Tbps of which 8 Tbps will be sellable [30, 32]. Seventy-eight of the satellites will be in a polar orbit at 99.5° inclination and are scheduled for launch 2022, while 220 satellites will be in inclined orbit of 50° and are scheduled for launch in 2023. The satellites will weigh about 800 kg and will have a design lifetime of 10 years, the longest proposed lifetime for LEO satellites. The satellites employ digital regenerative processing and digital phased arrays with optical intersatellite links [33].

A Phase 1 satellite was launched on a polar satellite launch vehicle (PSLV) operated by the Indian Space Research Organization (ISRO). The spacecraft was built by Surrey Satellite Technology Ltd., United Kingdom, a world leader in small satellites and part of the Airbus Defence and Space group [34].

Perhaps the most ambitious megaconstellation to date is the Starlink constellation operated by SpaceX. The original constellation included 12,000 satellites in three subconstellations, described as follows:

- The first includes 1,584 satellites operating in the Ku-band and Ka-band at an altitude of 550 km;
- The second includes 2,825 Ku-band and Ka-band spectrum satellites at 1,110 km;
- The third includes 7,500 V-band satellites at 340 km.

This was rationalized by putting the second subconstellation between 540 and 570 km altitude with orbit inclinations of 53.2°, 70.0°, and 97.6° and reducing the number of satellites by one [35]. This puts the number of satellites operating in the Ku-band and Ka-band to 4,408. Over 60 test satellites have been launched in two launch batches including the so called V0.9 satellites. At the time of writing, there have been 653 satellites launched as shown in Table 8.4, and of course, launches will continue.

These satellites, which weigh 260 kg, are of a flat design, allowing them to be stacked on each other for efficient Falcon 9 launching. Some satellites

Table 8.4
Launches of the Starlink Satellites at the Time of Writing [36, 37] (with All Satellites Having a Targeted Altitude of 550 km and an Inclination of 53°

Launch Date	Launcher	Flight Number	Number of Satellites
November 11, 2019	Falcon-9 v1.2 (Block 5)	75	60
January 7, 2020	Falcon-9 v1.2 (Block 5)	78	60
January 29, 2020	Falcon-9 v1.2 (Block 5)	80	60
February 17, 2020	Falcon-9 v1.2 (Block 5)	81	60
March 18, 2020	Falcon-9 v1.2 (Block 5)	83	60
April 22, 2020	Falcon-9 v1.2 (Block 5)	84	60
June 4, 2020	Falcon-9 v1.2 (Block 5)	86	60
June 13, 2020	Falcon-9 v1.2 (Block5)	87	58
August 7, 2020	Falcon-9 v1.2 (Block 5)	90	57
August 18, 2020	Falcon-9 v1.2 (Block 5)	91	58
September 3, 2020	Falcon-9 v1.2 (Block 5)	93	60

are launched directly in their intended orbit while others go through an orbit raise phase and may spend some time in the parking orbit of 380 km. Orbit raise uses the ion-propulsion system of the satellites. The satellites have a single solar array and employ digital payload technology using four phase arrays. The satellites are equipped with optical intersatellite links, star trackers, and autonomous collision-avoidance systems [38, 39]. These subconstellations have been designed so that the minimum terminal elevation is 40° [35]. SpaceX has been criticized by astronomers for the light pollution that the satellites create through reflection and has thus been working on ways to reduce its albedo, including the reduction of the original second subconstellation altitude and other techniques to reduce the reflection from the array antenna. The satellites are being manufactured by SpaceX with a current throughput of 120 satellites a month [40].

There are also 7,500 satellites that are planned to operate in the V-band at the low altitude of 340 km. The aerodynamic drag is considered to be appreciable and the satellite propulsion system will be used to compensate, which increases the demand on the spacecraft resource.

SpaceX has recently filed for an additional 30,000 satellites to build a constellation of 42,000 satellites [41].

Kuiper is another constellation proposed by Amazon. The system was filed with the FCC in 2019 [42]. Table 8.5 details the constellation, which is designed to provide seamless coverage between 56°S and 56°N. The system operates in the Ka-band and covers 95% of the world's population. Amazon has invested in a research and development facility in Redmond, Washington, and the first satellites are expected next year. Amazon and Blue Origin were founded by Jeff Bezos so that although its launcher, New Glenn, has not yet flown, it will be a good candidate for launching Kuiper satellites. The first satellites are expected to be launched in 2021.

Table 8.5
The Kuiper Constellation

Altitude	Inclination	Planes	Number of Satellites per Plane	Number of Satellites
630 km	51.9°	34	34	1,156
610 km	42.0°	36	36	1,296
590 km	33.0°	28	28	784
	Total	98		3,236

The history of LEO constellations is dotted with financial issues. The last three systems discussed are valued at $3 billion for Telesat LEO and $10 billion each for Starlink and Kuiper. This is roughly two or three magnitudes higher than a GEO HTS system. The revenues are, of course, also expected to be higher, as they potentially have global reach. However, operating a system globally and marketing a system globally does not come without challenges.

The coverage of LEO constellations is often ubiquitous for all longitudes. However, the demand over the oceans is not the same as the demand over land. Since 70% of the Earth's surface is ocean, at a simplistic level, for a global system there is a high demand for 30% of the system and a low demand for 70% of the system. Of course, not all land has the same demand. Nonetheless, as people expect to be connected wherever they are, the demand for broadband capacity continues to increase both on land, at sea, or in the skies.

The satellites are generally smaller but there are many of them, which has led the space industry to rethink satellite production. Outside constellations, the production runs of satellites have been typically limited to a handful and are more often one-off orders. Having to produce hundreds or even thousands of satellites is considerably different.

Traditional satellite operations have managed tens to a few hundreds of satellites often with different missions. Beside the satellite operations that ensure that the satellites are wherever they should be, there is a significant network management layer that ensures that the capacity is where it should be. There also needs to be orchestration in the hand-over between satellites so that users maintain their connectivity irrespective of which satellite is in use. The level of complexity is beyond anything that has been done.

Launching such a number is a challenge that has led to the concept of batch launching, the most significant being Starlink with launches in batches of up to 60 satellites. It is not surprising that both SpaceX and Kuiper have in-house access to launchers. The lower cost of launching has obviously been a positive factor for the business case.

LEO satellites have a lifetime that is a half to a third that of a GEO satellite. This has led the space industry to reconsider reliability requirements. The component and equipment requirements for a shorter lifetime are less demanding in their failure rate requirements, which can help in two ways: reducing the cost of a given function or pushing for more functionality. The latter has led to the use of advanced digital payload techniques.

The business model is also evolving [43]. The revenue for traditional satellite operators has been based on satellite usage. Now, however, increasingly, the revenue comes from the content. Often services are bundled together with the objective of increasing the overall revenue rather than looking at the business cases of services independently. Thus, satellite connectivity becomes one element in the business case at a bigger level with a longer-term vision.

The management of the system CAPEX and OPEX involved the higher frequency of replenishment of satellites, the handling of the network management layer, the operational issues of a global system with the required national operating rights, the marketing effort necessary in the different countries, and the potentially slower ramp-up in the system utilization are some of the challenges that the business case must incorporate. It will certainly be very interesting to see how the future of LEO unfolds.

References

[1]   NASA's Global Change Master Directory, Ancillary Description Writer's Guide, 2008, https://web.archive.org/web/20100528095529/http://gcmd.nasa.gov/User/suppguide /AD_Guide_2008.pdf.

[2]   NASA Safety Standard: Guidelines and Assessment Procedures for Limiting Orbital Debris, NASA Headquarters; Office of Safety and Mission Assurance; Washington, D.C., United States, Report Number: NSS-1740.14, August 1, 1995.

[3]   Steering Group and Working Group 4, IADC Space Debris Mitigation Guidelines, Inter-Agency Space Debris Coordination Committee, IADC Action Item number 22.4, IADC-02-01 Revision 1 September 2007.

[4]   ESA, Types of orbits, 30 March 2020, https://www.esa.int/Enabling_Support/Space _Transportation/Types_of_orbits.

[5]   ECSS-E-ST-10-04C Rev. 1 Working Group, Space Engineering, Space Environment, European Cooperation for Space Standardisation, Document Number: ECSS-E-ST-10-04C Rev. 1, ESA-ESTEC, Requirements & Standards Division, Noordwijk, The Netherlands, June 15, 2020.

[6]   Vette, J. I., "The AE-8 Trapped Electron Model Environment," *NSSDC/WDC-A-R&S Report 91-24*, NASA-GSFC, 1991.

[7]   Sawyer, D. M., and J. I. Vette, "AP8 Trapped Proton Environment for Solar Maximum and Solar Minimum," *NSSDC WDC-A-R&S Report 76-06*, NASA-GSFC, 1976.

[8]   Kosmonavtika, Les satellites Meridian (in French), 20 February 2020. https://www .kosmonavtika.com/satellites/meridian/meridian.html.

[9]   Geyer, M., "Earth-Referenced Aircraft Navigation and Surveillance Analysis," Project Memorandum, *Technical Report: DOT-VNTSC-FAA-16-12*, prepared for Federal Aviation Administration, Wake Turbulence Research Office, publisher John A. Volpe National Transportation Systems Center (U.S.), June 2016.

[10] Aerospace Technology, Globalstar Communication Satellites, https://www.aerospace -technology.com/projects/globalstar/#:~:text=SATELLITE%20CONSTELLATION

&text=The%20satellites%20are%20placed%20in,satellites%20in%20orbit%20as%20 spares.

[11] Iridium website, Iridium Satellite Communications, Factsheet, file:///C:/Users/htfen/ Downloads/FS_Iridium%20Corporate%20Overview_Fact%20Sheet_030220.pdf.

[12] Harms, J., The Orbcomm Experience, https://artes.esa.int/sites/default/files/1_The _Orbcomm_Experience.pdf.

[13] Ahmed, A., Orbcomm Satellite System, Faculty of Engineering, Cairo University Satellite Communication Course, published on June 2, 2014, https://www.slideshare .net/akrahm92/orbcomm-satellite-system.

[14] Barnett, R., "O3b—A Different Approach to Ka-Band Satellite System Design and Spectrum Sharing," *ITU Regional Seminar for RCC countries on Prospects for Use of the Ka-band by Satellite Communication Systems*, Almaty, Kazakhstan, September 5–7, 2012.

[15] Arianespace VS05 Launch kit, A Batch Launch for the O3b Constellation, https:// www.arianespace.com/wp-content/uploads/2015/10/VS05-O3b-launchkit-EN.pdf.

[16] Arianespace VS08 Launch kit, A Second Launch for the O3b Constellation, https:// www.arianespace.com/wp-content/uploads/2015/09/VS08-launchkit-GB.pdf.

[17] Arianespace VS10 Launch kit, A Third Launch for the O3b Constellation, https:// www.arianespace.com/wp-content/uploads/2015/09/VS10-launchkit-EN.pdf.

[18] Arianespace VS18 Launch kit, Arianespace Flight VS18—A Fourth Launch for the Operator SES and Its O3b Constellation, https://www.arianespace.com/wp-content /uploads/2018/10/DDP-VS18-final-GB.pdf.

[19] Arianespace VS22 Launch kit, Arianespace Flight VS22: A Fifth Launch for the Operator SES And Its O3b Constellation, https://www.arianespace.com/wp-content /uploads/2019/04/VS22-launchkit-EN-2.pdf.

[20] SES press release, "SES Opens New Era in Global Connectivity with O3b mPOWER," September 11, 2017.

[21] Boeing press release, "Boeing to Build Four Additional 702X Satellites for SES's O3b mPOWER Fleet," El Secundo, August 7, 2020.

[22] SES, Engineering Freedom, Redefining Network Services with O3b mPOWER, Insight SES, Betzdorf, Luxembourg, August 2019.

[23] Lloyd's satellite constellations, Overview: Teledesic, http://personal.ee.surrey.ac.uk /Personal/L.Wood/constellations/teledesic.html.

[24] Global Telecoms Insight: Teledesic, http://www.mobilecomms-technology.com /projects/teledesic/#:~:text=Its%20flexibility%20is%2C%20of%20course,planes%20 each%20with%2024%20satellites.

[25] OneWeb press release, "OneWeb Makes History as First Launch Mission Is Successful," London, February 28, 2019.

[26] OneWeb press release, "OneWeb Successfully Launches 34 More Satellites into Orbit," London, March 21, 2020.

[27] OneWeb press release, "OneWeb Successfully Launches 34 More Satellites into Orbit in Second Launch of 2020," London, February 21, 2020.

[28] OneWeb press release, "OneWeb Files for Chapter 11 Restructuring to Execute Sale Process," London, March 27, 2020.

[29] OneWeb press release, "OneWeb Seeks to Increase Satellite Constellation Up to 48,000 Satellites, Bringing Maximum Flexibility to Meet Future Growth and Demand," London, May 27, 2020.

[30] OneWeb press release, "OneWeb Announces HMG and Bharti Global Limited consortium as Winning Bidders in Court-Supervised Sale Process," July 3, 2020.

[31] OneWeb statement, July 27, 2020, https://www.oneweb.world/media-center/oneweb-files-for-chapter-11-restructuring-to-execute-sale-process.

[32] Henry, C., "Telesat Preparing for Mid-2020 Constellation Manufacturer Selection," *Space News*, May 1, 2020, https://spacenews.com/telesat-preparing-for-mid-2020-constellation-manufacturer-selection/.

[33] Telesat Website, Telesat LEO https://www.telesat.com/leo-satellites/.

[34] Telesat press release, "Telesat Begins Deploying Its Global Low Earth Orbit (LEO) Constellation with Successful Launch of Phase 1 Satellite," Ottawa, Canada, January 12, 2018.

[35] Application of SPACE EXPLORATION HOLDINGS, LLC to FCC, Application for Modification of Authorization for the SpaceX NGSO Satellite System, 17 April 2020, https://fcc.report/IBFS/SAT-MOD-20200417-00037/2274315.pdf.

[36] Wikipedia: Starlink, https://en.wikipedia.org/wiki/Starlink.

[37] Wikipedia: List of Falcon 9 and Falcon Heavy Launches, https://en.wikipedia.org/wiki/List_of_Falcon_9_and_Falcon_Heavy_launches.

[38] SpaceX website, High Speed Internet across the Globe, April 28, 2020 https://www.starlink.com/.

[39] SpaceX website, Starlink Discussion National Academy of Sciences, 28 April 2020 https://www.spacex.com/updates/starlink-update-04-28-2020/.

[40] Sheetz, M., "SpaceX Is Manufacturing 120 Starlink Internet Satellites per Month," CNBC Website, 10 August 2020, https://www.cnbc.com/2020/08/10/spacex-starlink-satellite-production-now-120-per-month.html.

[41] Application for Approval for Orbital Deployment and Operating Authority for the SpaceX Gen2 NGSO Satellite System, to the FCC, May 26, 2020, https://fcc.report/IBFS/SAT-LOA-20200526-00055/2378669.pdf.

[42] Application of Kuiper Systems LLC for Authority to Launch and Operate a Non-Geostationary Satellite Orbit System in Ka-band Frequencies, Application for Authority

to Launch and Operate a Non-Geostationary Satellite Orbit System In Ka-Band Frequencies, July 4, 2019, https://docs.fcc.gov/public/attachments/FCC-20-102A1.pdf.

[43] Daehnick, C., et al., "Large LEO Satellite Constellations: Will It Be Different This Time?," McKinsey & Company, Article, May 2020, https://www.mckinsey.com/~/media/McKinsey/Industries/Aerospace%20and%20Defense/Our%20Insights/Large%20LEO%20satellite%20constellations%20Will%20it%20be%20different%20this%20time/Large-LEO-satellite-constellations-Will-it-be-different-this-time-VF.pdf.

9

HTS Systems in the 5G Ecosystem

9.1   Introduction

5G promises huge improvements in mobile communications and covers areas of communications that the previous generations did not include, namely massive machine-type communications and low-latency high-reliability communications or industrial IoT. 5G, a network of networks that provides the best possible communications wherever one is, embraces satellite communications as having an important role to play to meet its objectives. 5G represents an evolution since it is the fifth generation and builds on the previous generations to push mobile communication to new frontiers. Moreover, it is a revolution because it goes well beyond mobile communications.

This chapter looks at the previous generations of mobile communications and then looks more closely at the objectives of 5G and the role of satellite communications in the 5G ecosystem.

9.2   1G, 2G, 3G, 4G

5G follows the first four generations of mobile communications, with 1G analog for voice and digital for signaling and all the other generations fully digital. The switch to digital took place with the advent of 2G, which introduced digital access schemes such as time-division multiple access (TDMA) and code-division multiple access (CDMA) [1, 2]. CDMA systems were mostly used in the Americas. Digital communications improved spectral efficiency and the quality of security and introduced SMS. In 1983, in Europe, the CEPT established the *Groupe Spéciale Mobile* (GSM) that brought 13 countries to develop a European standard for an open digital cellular system for transmitting mobile voice and data services. The group later maintain the acronym for Global System for Mobile Communications. The system was to employ a combination FDMA and TDMA. In 1989, the group was transferred to ETSI. In 1986, the heads of state of the European Union endorsed the project and adopting the 900-MHz spectrum. In 1989, the 1,800-MHz spectrum was included. This led to an open standard that was adopted in many countries outside of North America. The first GSM call was made in Finland 1991. By 1998, the standard was adopted by more than 100 countries, and there were more than 100 million users [3]. In 2002 the GSM was adopted by 75% of the worldwide networks [4].

The next generation to come was 3G, whose requirements were developed by the ITU as the International Mobile Telecommunications-2000 (IMT-2000) specifications [5]. This includes the following six technologies:

- CDMA direct spread;
- CDMA multicarrier;
- CDMA time-division duplexing (TDD);
- TDMA single-carrier;
- FDMA/TDMA;
- Orthogonal frequency division multiple access (OFDMA) TDD wireless metropolitan area network (WMAN).

These radio interfaces support the features and design parameters of IMT-2000, including the capability to ensure worldwide compatibility, international roaming, and access to high-speed data services such as video calling, streaming, e-mails, Internet, and video. Thus, 3G is not one standard but a family of interoperable standards. Two main standards emerged, the Universal Mobile Telecommunications System (UMTS) and CDMA-2000. The latter was favored in North America and South Korea. UMTS is probably the best-known of the technologies and was standardized by the 3GPP and used primarily in Europe, Japan, and China. UMTS operators can operate their 3G services from anywhere in the world using satellite and land-based infrastructure. UMTS employs direct-sequence or wideband CDMA (W-CDMA) radio access technology to enhance the spectral efficiency and bandwidth utilization for mobile network operators. It uses channels with bandwidths of 10 MHz, 5 MHz, and 1.6 MHz with corresponding chip rates of 7.68 Mchip/s, 3.84 Mchip/s, and 1.28 Mchip/s, respectively, in combinations of time and frequency division duplex. Initially specified for the 2.1-GHz band, it was then opened to several bands from 800 MHz to 3.5 GHz [6]. The first commercial network was launched in Japan in 2001.

The fourth generation introduced broadband to mobile communications, connected anywhere anytime. The requirements for 4G are defined by the ITU as IMT Advanced [7]. The emphasis was on enhanced data rates and mobility. Aggregate cell spectral efficiencies range from 1.1 bps/Hz to 3.0 bps/Hz and 0.7 bps/Hz to 2.25 bps/Hz for the downlink and the uplink respectively with peak figures of 15 bps/Hz and 6.75 bps/Hz. Enhanced mobility also brings challenges, especially mobility across network boundaries, thus bringing to the foreground location management and handover management, reducing dropouts as a mobile terminal moves from one cell to another. Mobility covers from stationary to 350 km/h. 4G is a convergence of networks to an integrated network that is IP-based to extend services to other platforms like computers and consumer electronics and *things* with substantial speeds, even when in motion. Two technologies emerged, namely long-term evolution (LTE) Advanced as standardized by the 3GPP/ETSI and WiMAX or IEEE 802.16 as standardized by the IEEE. The original LTE was intended to be an enhancement of the 3G UMTS, but LTE Advanced [8] was conceived for 4G. LTE Advanced includes features like 4x4 MIMO, 256 QAM, and carrier aggregation, which allows groupings of up to five channels of bandwidths between 1.4 and 20 MHz to be made available to a single user. These techniques contribute significantly more capacity to the user [9].

Table 9.1 briefly compares the 2G, 3G, and 4G systems.

Table 9.1
Summary Comparison of 2G, 3G, and 4G

Generation	2G	3G	4G
Start	1980	1990–2000	2000
Deployment	1990	2002	2010
Data rate	64 kbps	2 Mbps	1 Gbps
Services	Digital voice, SMS	Enhanced streaming, video conferencing, web browsing, IPTV	Enhanced streaming, VoIP, HD mobile TV
Multiplexing	TDMA, FDMA	CDMA	CDMA

9.3   5G

5G is the first generation of mobile networks to push broadband mobile communications far beyond the previous generations. It also addresses a wide range of needs of vertical industries by design. It provides massive connectivity for everything from human-held smart devices to sensors and machines. Moreover, 5G can support critical machine communications with low latency and ultra-high reliability. The vision of the 5G system was expressed by the ITU in a report on the future development of IMT [10]. 5G is also referred to by the ITU as IMT Advanced. As the name implies it is considered as an extension of IMT providing diverse services in three use cases. These are enhanced mobile broadband (eMBB), ultrareliable and low-latency communications (URLLC), and massive machine type communications (mMTC). The requirements are given in [11] and outlined as follows:

- Peak data rate: The maximum ideal achievable data rate for a given user. Peak data rate is defined for a single mobile station. For a given band, it is related to the peak spectral efficiency in that band. Communications may occur over a number of bands so that the rate is the aggregate rate taking all the bands employed. The minimum requirements are 20 Gbps and 10 Gbps downlink and uplink, respectively. This is applicable for eMBB.
- *Peak spectral efficiency:* The peak data rate in relation to the bandwidth used. The minimum requirements are 30 bps/Hz and 15 bps/Hz for the downlink and the uplink, respectively. This is applicable for eMBB.
- *User-experienced data rate:* The data rate for a given user for 95% of the time and aggregated in the case of multiband utilization. The minimum requirements in dense urban conditions are 100 and 50 Mbps for the downlink and the uplink, respectively.
- Fifth percentile user spectral efficiency: This is the spectral efficiency in bps/Hz for a given user for 95% of the time. This estimation is based on the number of received error-free bits of user, the active session time, and the channel bandwidth. This is applicable for eMBB. Table 9.2 lists the minimum requirements under the various conditions.
- Average spectral efficiency: This is the aggregated over all users. This requirement is applicable in the eMBB scenario. Table 9.3 lists the requirements.
- *Area traffic capacity:* The total traffic throughput served per geographic area, specified for a downlink in an indoor hotspot as 10 Mbps/m in the eMBB context.

Table 9.2
Minimum Requirements for 5th-Percentile User Spectral Efficiency for Indoor Hotspot, Dense Urban, and Rural Conditions

Environment	Downlink (bps/Hz)	Uplink (bps/Hz)
Indoor Hotspot	0.300	0.210
Dense Urban	0.225	0.150
Rural	0.120	0.045

- *Latency:* The time in the radio network from when the source sends a packet to when the destination receives it, is defined as 1 ms for eMBB and 4 ms for URLLC. Control plane latency refers to the transition time from a most "battery-efficient" or standby state to the start of continuous data transfer and is defined as 20 ms for both eMBB and URLLC.
- *Connection density:* The total number of devices fulfilling a specific QoS per unit area, defined as 1,000,000 devices per km^2 for mMTC applications.
- *Device energy efficiency:* The capability of the device modem of a given technology to minimize the power consumed in relation to the traffic characteristics. It is a function of the sleep ratio. The network energy efficiency is also defined. These requirements are applicable in the eMBB case.
- *Reliability:* This relates to the capability of transmitting a given amount of traffic within a predetermined time duration with a given probability. The requirement is 99.999% probability of transmitting of 32 bytes within 1 ms at the coverage edge for the URLLC case.

Table 9.3
The Minimum Requirements for Average Spectral Efficiency for Indoor Hot Spot, Dense Urban, and Rural Conditions

Environment	Downlink (bps/Hz/Transmit Receive Point)	Uplink (bps/Hz/ Transmit Receive Point)
Indoor Hotspot	9	6.75
Dense Urban	7.8	5.4
Rural	3.3	1.6

Table 9.4
The Specified Data Rates at the Given Speeds

Environment	Normalized Traffic Channel Link Data Rate (bps/Hz)	Mobility (km/h)
Indoor hotspot	1.50	10
Dense urban	1.12	30
Rural	0.80	120
	0.45	500

- *Mobility:* The maximum speed at which a defined QoS can be achieved, namely stationary, pedestrian at up to 10 km/h, vehicular at speeds of 10 km/h to 120 km/h, and high-speed vehicular between 120 km/h and 500 km/h. The last figure covers mainly high-speed trains. The main application is eMBB. Table 9.4 lists the specified data rates in these conditions.
- *Mobility interruption time:* The duration in which a user terminal cannot exchange user plane packets with any base station. This includes time for radio access, resource control, and signaling and should be zero. This is defined for both eMBB and URLLC cases.
- *Bandwidth:* The aggregated system bandwidth. The bandwidth may be supported by single or multiple RF carriers. The minimum requirement for is 100 MHz.

There is no single spectrum that would meet all these requirements, and consequently, 5G operates on three different bands, described as follows:

- *The low-band spectrum:* The sub-1-GHz band has been extensively used by the previous systems. It offers good area coverage and wall penetration but is limited in spectrum, and peak speeds are limited to about 100 Mbps. This spectrum allows ubiquitous coverage within the designated service area. The expected range from the tower would be around 15 km. Table 9.5 shows the spectrum identified by the ITU for IMT at the low band [12, 13]. Notice that the total of 510 MHz of spectrum has been identified at the World Radiocommunications Conference of 2015 (WRC-15). The applicability is estimated by the number of countries in the identification. There are 193 member states in the ITU.

Table 9.5
Spectrum Identified by the ITU for IMT and 5G Low-Band

Band				
Start	**End**	**Bandwidth**	**Applicability**	**WRC**
450 MHz	470 MHz	20 MHz	100%	WRC-15
470 MHz	698 MHz	228 MHz	4%	WRC-15
698 MHz	960 MHz	262 MHz	100%*	WRC-15
	Total	510 MHz		

*Not all the spectrum in every region.

- *The medium-band spectrum:* The sub-6-GHz band (or more precisely the 1–6-GHz band) is more limited in wall penetration, but more spectrum is available to deliver peak speeds of 1 Gbps. The expected range would be about 3 km. Massive MIMO and beamforming can be used in improve penetration since the required signal level is concentrated toward the wanted user to establish the communication link more efficiently. The sub-6-GHz band allows improved performance over larger areas. Table 9.6 lists the spectrum identified by the ITU for the mid band [12, 13]. Like the low band, the spectrum was specifically identified for IMT at WRC-15. A total of about 1.5 GHz is

Table 9.6
Spectrum Identified by the ITU for IMT and the 5G Medium Band

Band				
Start	**End**	**Bandwidth**	**Applicability**	**WRC**
1,427 MHz	1,518 MHz	91 MHz	100%*	WRC-15
1,710 MHz	2,025 MHz	315 MHz	100%	WRC-15
2,110 MHz	2,200 MHz	90 MHz	100%	WRC-15
2,300 MHz	2,690 MHz	390 MHz	100%	WRC-15
3,300 MHz	3,400 MHz	100 MHz	23%	WRC-15
3,400 MHz	3,600 MHz	200 MHz	87%	WRC-15
3,600 MHz	3,700 MHz	100 MHz	2%	WRC-15
4,800 MHz	4,990 MHz	190 MHz	2%	WRC-15
	Total	1,476 MHz		

*Not all the spectrum in every region.

listed. The applicability is estimated by the number of countries in the identification. The spectrum between 3.4 and 3.6 GHz was originally allocated to satellites designated as part of the C-band.

- *The high-band spectrum:* The above-6-GHz spectrum is what allows the impressive performance often quoted for 5G, but it is available over relatively small areas as these frequencies suffer higher attenuations due to obstacles and foliage. This band is specific to 5G. The performance is typically available up to about 500m from the tower. Table 9.7 shows the spectrum identified by the ITU for the high band and adds up to 17.25 GHz of spectrum, which was identified recently at WRC-19 [14]. The applicability is estimated by the number of countries in the identification.

Note that speeds are a function of distance from the tower so that over a given area the minimum performance increases with the tower density. Also, the range can vary considerably according to the natural terrain and obstacles, which, for example, can produce significant differences whether the cell phone is inside a building or outside. It is also a trade-off that the operator uses to program investment against speed. A network that is dense with towers over a given area reduces the distance of devices from the tower and increases speed.

A simplistic analysis demonstrates that the number of towers over a given area increases dramatically with the frequency of the band due to decreasing range. Using the numbers given above and with the caveat that these can only be indicative because of the geographical variability and the trade-off involved, for a given area going from low band to medium band, the number of towers required would increase by a factor of 25. This due to the fact that ranges for

Table 9.7
Spectrum Identified by the ITU for the 5G High Band

Band				
Start	End	Bandwidth	Applicability	WRC
24.25 GHz	27.50 GHz	3,250 MHz	100%	WRC-19
37.00 GHz	43.50 GHz	6,500 MHz	100%	WRC-19
45.50 GHz	47.00 GHz	1,500 MHz	27%	WRC-19
47.20 GHz	48.20 GHz	1,000 MHz	54%	WRC-19
66.00 GHz	71.00 GHz	5,000 MHz	82%	WRC-19
	Total	17,250 MHz		

the low and medium bands are given as 15 km and 3 km respectively. The required number of towers for a given area thus increases to the square of the ratio of the ranges. Similar going from low band to high band the factor is 900. This underlies a frequent criticism of 5G: that the capital outlay could be huge, but performance has a cost. It also explains why some operators are getting cold feet about the high band. Of course, operators do not have to offer the same performance over their service areas but then we enter a different digital divide of the well-served and the relatively underserved, a sort of second class of customer.

Although the three bands are identified for 5G, there is no obligation for operators to provide services in all three bands. The extension to the high-band and the expansion over the required service area is likely to be a major economic consideration.

9.4 What Role Can HTS Play?

The importance of satellite system as a complementary component to the terrestrial component was recognized by the ITU in [15, 16] in the 3G days when it was deemed that satellites could complement the terrestrial system with the introduction mobile satellite units and/or mobile hybrid units. At the time, bit rates of up to 144 kbps were considered, with 16 kbps being the limit for hand-held units. The system was based on a direct satellite link using bands between 1 and 3 GHz. The speeds specified for 5G make direct mobile satellite communications challenging from the link budget and dimensioning points of view. This is particularly the case for geostationary satellites. This is a dimensioning issue, and considering that the main thrust of mobile communications is based on hand-held devices, it is not likely that direct device to satellite communication will be pursued for mainstream mobile communications. The size of a typical cell phone is considerably smaller than that of a satellite phone.

This does not mean that there is no role for satellites in 5G. The role is more on the infrastructure side [17, 18]. 5G has often been described as a network of networks, and the satellite network is one of them. With 5G network slicing, the network architecture employs virtualized and independent logical networks on the same physical network infrastructure. Each network slice is an end-to-end network that allows the optimization of a service. Use cases have been defined for 5G, but the list is not exhaustive, and the system is intended to remain open to future services. 5G will have the role of supporting services in an economic and sustainable fashion and not on an ad hoc basis as has often been the case with satellite communications in the past. Standardization will help the integration of satellites as another network building block when they offer the best solution and thereby avoid unnecessary customization. It should be noted that standardization is intended to refer to the required interfaces and not to technologies. Hybridization in 5G ensures that the technologies can coexist and complement each other to deliver services that may not be possible otherwise.

The EMEA Satellite Operators Association (ESOA) and the Global Satellite Coalition recognizes four use cases [19] where satellites can contribute to the 5G system. The four use cases are also given in a white paper [20] by the 5G Infrastructure Public Private Partnership (5G PPP), which is a joint initiative between the European Commission and the European ICT industry (ICT manufacturers, telecommunications operators, service providers, SMEs, and research institutions). All these use cases rely on satellites as a transport network linking the access network to the core network.

Satellite backhauling is not new for mobile communications. The base station or tower forms the central node of a star network or the access network with the active devices in the cell. The backhaul link connects the central node to the core network. Several technologies are available such as wire, fiber, microwave, and satellite.

Figure 9.1 shows a 5G network architecture where a number of towers manage communications with the devices in their cell, each forming a sub-network or a radio access network. In Figure 9.1(a) the tower communicates with the operator core network through terrestrial means such as a fiber or a microwave link, thus using the terrestrial component. In Figure 9.1(b) the tower communicates to the core network through a satellite link. The interface at the tower between the 5G section and the satellite section could be baseband so that the satellite link could use the HTS interface, exploiting standard equipment with an optimized satellite capacity usage and using the satellite infrastructure minimizing both CAPEX and OPEX. In all cases, communications within the access network is standard as defined in 5G. A user can utilize the same device, and whether the tower communicates with the core network through a terrestrial or a satellite network is transparent to the user.

It is known that satellites, especially geostationary ones, incur intrinsic round-trip delays that are not compatible with the 5G response requirements. The round-trip delay for GEO satellites is about 250 ms, which reduces to some 4 ms for LEO satellites. However, satellites, including geostationary satellites, can contribute to better the latency of a 5G system, even helping to meet the

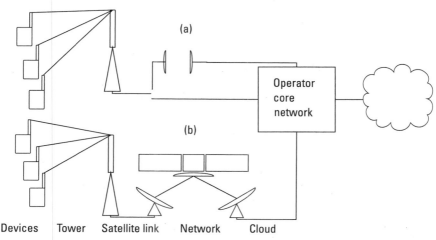

Devices   Tower   Satellite link   Network   Cloud

Figure 9.1   5G backhauling over satellite. (a) shows a tower with its subnetwork connected to the core network via wire, fifer or or a microwave link, while (b) show a similar subnet where the tower is connected to the core network through a satellite link. In both cases, this is transparent to the end user.

1-ms requirement. One of the factors that contribute to latency in data access is the distance within the network between the device and the location of the data. Satellites, in general, excel in broadcast, thus bringing the data closer to users through distributed storage, which sends the data closer to the devices to reduce latency when the data is requested. Delivering data to the edge and caching it in local storage, as close as possible to potential users, makes content readily and rapidly available. This is the principle of edge computing. HTSs and VHTSs can transfer even more data with specific profiles to take into account the local requirements.

HTS systems can also do an excellent job to repatriate data collected from IoT devices. Devices within the access network send their data to the tower, and the aggregated data is relayed via satellite to the operator core network for delivery and processing. Chapter 3 discusses a similar system.

Reference [21] compares the total cost of ownership for a backhaul system employing microwave and satellite technologies. The white paper concludes that HTS systems offer the most cost-effective capacity for a backhaul solution. This is expected to improve further with larger HTS systems. Terrestrial backhaul becomes less cost-effective as the region becomes more rural. CAPEX and maintenance costs are a lower percentage of the overall expense

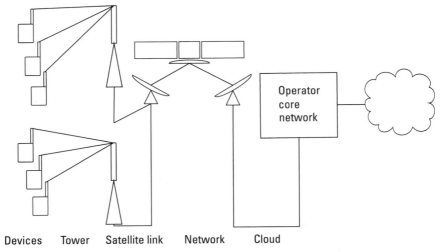

Devices   Tower   Satellite link   Network   Cloud

Figure 9.2   5G trunking over satellite.

in a satellite network. As HTS systems become more prolific, both these costs are expected to drop further. Satellite infrastructure is portable since it can be reused wherever it is needed within the satellite service area. Time to market is shorter, since standard equipment can be used, and the associated licensing is on a country basis and not specific to the site.

Trunking involves sharing capacity among a number of towers so that the number of devices sharing the capacity is larger and the total bandwidth requirements are larger. The towers are networked to a central node. When using satellites, the central node is linked to the operator core network via satellite, thus forming a network of access networks (Figure 9.2). Trunking offers a higher level of aggregation, a higher level of statistical multiplexing, and a better economy of usage of the satellite resource and ground satellite stations. Cache or storage and edge computing may also be available at this site for access of the data through the towers associated with the central node, thus enhancing latency of the 5G system.

Applications of this use case could include a remote agglomeration where the required service area is covered by a terrestrial 5G network but the connection to the operator core network is performed via satellite. In trunking, the capacity requirements are generally higher than those of backhauling. This means that a trade-off may exist between using (perhaps a small number of) standard satellite terminals and more performing larger terminals. Standard terminals could be faster to install and entail less CAPEX but potentially result is a higher OPEX because of lower link efficiencies (using lower-order modulation and more coding). Larger terminals are obviously more expensive but use the link more efficiently and could reduce OPEX. The choice will depend on the capacity requirements and the required inception of service among other parameters in the equation.

Disaster relief could be used when the infrastructure is disabled to rapidly restore service through transportable equipment. For example, in 2016, Hurricane Mathew was the most damaging hurricane to hit the Caribbean for a decade. It was particularly damaging in Haiti where the communication system was badly hit. This not only added to the difficulties of the local community but also hampered the relief operation. Télécoms Sans Frontières, an NGO specializing in emergency telecommunications, used Eutelsat 117 West A [22] to restore telecommunications on the island. Although this example did not involve 5G, it is a demonstration of the use case that could be included in the 5G architecture. Such events are not limited to the Caribbean but are occur elsewhere. For example, flooding is happening more frequently in the United States, France, and the United Kingdom, with its ensuing impact on the local infrastructure.

First responders may require 5G services in remote areas where terrestrial infrastructure may not exist. A rapid network can be established that offers a wide range of services, such as IoT, to the rescuers so that their position can be monitored for safety reasons. For example, IoT sensors could be used for environmental monitoring, such as temperature, over a given area, with voice communications available for coordination of the operation. Video data using drones can also be used for further assessment of the overall situation, and the evolution over time with the detail can be decided on-site for maximum agility and effectiveness. This can be relayed back to the base for expert analysis and operation coordination. Video data can also be exchanged with the medical support team to facilitate and enhance the performance of action. The network can be based on 5G where most of the functions are already available on site and at the medical centers. Such scenarios could include an aircraft crash in a remote area or a fire in a forest. In a forest fire, sensors could send localized information on temperature and humidity to monitor the area and allow responders to act with greater speed and with the required force and efficiency, making best use of the available resources.

An extension of this use case could also work in military operations where the network could initially be established for ISR purposes, which would require minimal effort to install and could even be left unmanned after installation to monitor the situation. The network could then grow and be adapted according to the evolution and the associated requirements. Eventually, trunking could be introduced to cover the evolving needs. In the end, the act of connecting a remote agglomeration or a remote theater share several issues.

Satellite trunking could also be used to augment the resilience of the 5G network as a backup and/or to absorb peaks to critical areas or remote islands or areas. Thus, under normal conditions, terrestrial means are employed to connect to the operator core network, but it this were to fail or if the peak requirement exceeded the capacity of the terrestrial means, the satellite link could be used.

Hybrid multiplay pushes the satellite connectivity to the premises, which could be a home or a building. Figure 9.3 illustrates this, showing a premises that can be considered as a small contained cell connected to the 5G network via satellite. HTSs and VHTSs are well suited for this purpose, as it capitalizes on the broadcast and multicast function of satellites. Video and TV contents along with other data can be delivered to the building to be stored and readily available when required. In a company building, the local cache can be pro- filed to the specific requirements. This is especially useful in an underserved area. The quality of experience can be greatly improved to become similar to that of a well-served area.

Communication on the moving platform is where satellites shine, as this is not really possible with terrestrial means or is, at best, considerably limited in coverage. The main applications are maritime and aeronautical but trains are also to be included. This can be considered as the mobile version of backhauling. Figure 9.4 shows a scheme in which the passengers use a local 5G access network on the moving platform that is linked to the core network via a satellite. The use of cache can enhance the user experience, as video and

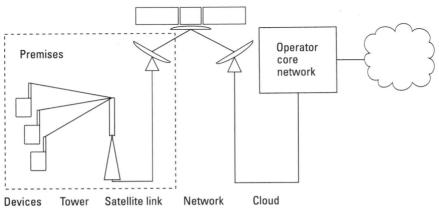

Devices   Tower   Satellite link   Network   Cloud

Figure 9.3   5G to the premises using satellites.

other content can be loaded when the aircraft is at the apron or the ship is in the port. This allows a large base of content data to be available locally, and what is not available in the cache can be loaded *on demand*. Of course, the user is not aware of what content originates from the local cache and what content is using the satellite to download. This use case is a prime example for satellite use as there is no other alternative, especially over the ocean where most of the long-haul aircraft, and many ships, are located.

Chapter 3 discusses aeronautical services, which are referred to as IFC, and maritime services, which can be divided into two groups, those used by passenger vessels and those used by cargo ships. Passenger vessels include cruise ships and ferries, while services on cargo ships include fleet management and crew welfare. The communication requirements for IFC and passenger vessels are growing markets. As people become increasingly accustomed to being connected, they expect the same services whether they are travelling for pleasure or for business on an aircraft or a passenger vessel.

Mobile connectivity already represents 6% of the revenue for Eutelsat [23], which has two satellites with specific payloads for this application. Eutelsat 172B was launched on 1 June 2017 and located at 172° East. Its mobility payload delivers 1.5 Gbps over the North Pacific for IFC. Eutelsat 10B is scheduled for launch in 2022 [24] to be located at 10° East. It will include two multibeam HTS Ku-band payloads: a high-capacity payload, covering the North Atlantic corridor, Europe, the Mediterranean basin, and the Middle East, offering significant throughput in the busiest air and sea traffic zones, and a second payload to extend coverage across the Atlantic Ocean, Africa, and the Indian Ocean. Eutelsat 10B will deliver 35 Gbps for maritime and IFC.

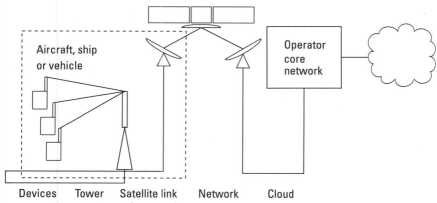

Figure 9.4   5G on a moving platform using a satellite.

Fleet management for cargo ships enhances the efficiency of the operation. A ship can make good use of information about port delays and meteorology information to optimize fuel consumption for the given route under the prevailing conditions. If the engine telemetry indicates a certain potential problem, the spare parts and the right personnel can be made available at the first port of call to ensure minimum downtime. Welfare communication helps the crew remain in touch with their families and friends. This eases their life at sea and can be used to attract crew.

5G can also be used to monitor containers on ships. For example, containers could have an IoT tag so that if they hold sensitive cargo, the environmental conditions can be continuously verified to ensure that the contents arrive in the best condition. The IoT tag can also provide the location of the container. This can be used for planning purposes as owners can track their assets, but it is also useful to know if the container is on the ship or lost at sea. Estimates on the number of containers lost at sea differ, and there are variations from year to year, but according to the World Shipping Council [25], 568 containers are lost on average each year at sea, excluding catastrophic events, and another 1,014 containers are lost on average at sea each year in catastrophic events, making a total of about 1,600. This can represent a considerable hazard to navigation, so knowing the position of lost containers or even the position where their contact was lost with the network on the ship can be useful to safety.

Trains can also offer passengers 5G services. When in urban areas, they can connect to the 5G network through terrestrial wireless means, and when they are in rural areas, they can employ an HTS connection to ensure that passengers enjoy the best quality of experience even when the train is going at 300 km/h. Similarly, passengers on long-distance buses may profit from such services, albeit on a smaller scale, since a French Train à Grande Vitesse (TGV), for example, may carry up to 500 passengers, while a bus transports an order of magnitude fewer passengers.

References

[1]   Brainbridge, "From 1G to 5G: A Brief History of the Evolution of Mobile Standards," https://www.brainbridge.be/news/from-1g-to-5g-a-brief-history-of-the-evolution-of-mobile-standards.

[2]   Devra, S., N. Bhandari, and K. Singh, "International Evolution of Cellular Network: From 1G to 5G," *Journal of Engineering and Techniques*, Volume 3 Issue 5, Sep.–Oct. 2017.

[3]   GSM Association, *Brief History of GSM & the GSMA*, https://www.gsma.com/aboutus /history.

[4]   Shneyderman, A., and A. Casati, *Mobile VPN: Delivering Advanced Services in Next Generation Wireless Systems*, Indianapolis, Indiana: Wiley Publishing, 2003.

[5]   Recommendation ITU-R M.1457-14, (01/2019), *Detailed specifications of the terrestrial radio interfaces of International Mobile Telecommunications-2000 (IMT2000)*, M Series, Mobile, radiodetermination, amateur and related satellite services, International Telecommunications Union, Geneva, Switzerland, 2019.

[6]   *Universal Mobile Telecommunications System (UMTS); Physical Layer—General Description* (3GPP TS 25.201 version 14.0.0 Release 14), ETSI TS 125 201 V14.0.0 (2017-04).

[7]   Report ITU-R M.2134, *Requirements related to technical performance for IMT-Advanced radio interface(s)*, International Telecommunications Union, Geneva, Switzerland, 2008.

[8]   *LTE: Requirements for further advancements for Evolved Universal Terrestrial Radio Access (E-UTRA) (LTE-Advanced)* (3GPP TR 36.913 version 15.0.0 Release 15), ETSI TR 136 913 V15.0.0 (2018-09).

[9]   Jeanette Wannstrom, *LTE-Advanced, 3PPP*, submitted on June 2013, https://www.3gpp .org/technologies/keywords-acronyms/97-lte-advanced.

[10]   Recommendation ITU-R M.2083-0, *IMT Vision—Framework and Overall Objectives of the Future Development of IMT for 2020 and Beyond*, M Series Mobile, Radiodetermination, Amateur and Related Satellite Services, International Telecommunications Union, Geneva, Switzerland, September, 2015.

[11]   Report ITU-R M.2410-0, *Minimum Requirements Related to Technical Performance for IMT-2020 Radio Interface(s)*, M Series Mobile, radiodetermination, amateur and related satellite services, International Telecommunications Union, Geneva, Switzerland, November, 2017.

[12]   Article 5, *Frequency Allocations*; Chapter 2, Frequencies; Volume 1, Articles; Radio Regulations, Edition of 2016, International Telecommunication Union, Geneva, Switzerland, ISBN-10 : 9261199976, ISBN-13 : 978-9261199975.

[13]   Restrepo, J., "Spectrum Allocation for 5G International Framework," *ITU Regional Economic Dialogue on Information and Communication Technologies for Europe and CIS (RED-2019) Regulatory and Economic Tools for a Dynamic ICT Market Place*, Odessa, Ukraine, October 30–31, 2019.

[14]   *World Radiocommunication Conference 2019 (WRC-19), Provisional Final Acts*, International Telecommunications Union, Sharm El-Sheikh, Egypt, 28 October–22 November 2019, https://www.itu.int/en/ITU-R/conferences/wrc/2019/Documents /PFA-WRC19-E.pdf.

[15]   Recommendation ITU-R M.818-2, *Satellite Operation within International Mobile Telecommunications-2000 (IMT-2000)*, International Telecommunications Union, Geneva, Switzerland, June 2003.

[16]   Recommendation ITU-R M.1850-2, *Detailed specifications of the Radio Interfaces for the Satellite Component of International Mobile Telecommunications-2000 (IMT-2000)*, M Series Mobile, Radiodetermination, Amateur and Related Satellite Services, International Telecommunications Union, Geneva, Switzerland, October 2017.

[17]   Fenech, H., "Satellite in the 5G Ecosystem," keynote address at the *9th EAI International Conference on Wireless and Satellite Systems (WiSATS 2017)*, Oxford, United Kingdom, September 14–15, 2017.

[18]   Agnelli, S., G. Benoit, and E. Weller, "Eutelsat Perspective on the Role of Satellites in 5G," *European Conference on Networks and Communications 2016 (EuCNC2016)*, Athens, June 27–30, 2016,

[19]   EMEA Satellite Operators Association, *Satellite Communication Services: An Integral Part of the 5G Ecosystem*, White Paper, January 2020.

[20]   5G PPP Architecture Working Group, *View on 5G Architecture*, White Paper, Version 3.0, DOI 10.5281/zenodo.3265031, February 2020.

[21]   Gilat Satellite Network Ltd., *Satellite Backhaul vs. Terrestrial Backhaul: A Cost Comparison*, White Paper, July 2015.

[22]   Eutelsat News, *Satellite Communications, an Alternative for the Haitian Population after Hurricane Matthew*, 27 October 2016,16:18 CEST, https://news.eutelsat.com/news/satellite-communications-an-alternative-for-the-haitian-population-after-hurricane-matthew-194272.

[23]   Third Eutelsat press release, "Quarter and Nine Month 2019–20 Revenues," 14 May 2020, 17:44 CEST, https://news.eutelsat.com/pressreleases/third-quarter-and-nine-month-2019-20-revenues-2999415.

[24]   Eutelsat press release, "Eutelsat Orders Eutelsat 10b Satellite for Inflight and Maritime Connectivity Services," 29 October 2019, 08:34 CET, https://news.eutelsat.com/press-releases/eutelsat-orders-eutelsat-10b-satellite-for-inflight-and-maritime-connectivity-services-2936959.

[25]   World Shipping Council, *Containers Lost At Sea—2017 Update*, 2017, http://www.worldshipping.org/industry-issues/safety/Containers_Lost_at_Sea_-_2017_Update_FINAL_July_10.pdf.

10

The Evolving Broadband Satellite Environment

10.1   Introduction

Chapters 1–9 discuss, for the most part, systems employing conventional HTS payloads. This is because most of the current HTS systems use such payloads. However, the situation is evolving toward the digital payload. There are a few examples of digital payloads in services like Inmarsat's first generation of Global Xpress GX1 to GX4. Others are in production, like SES's O3b mPOWER satellites and Inmarsat's second generation of Global Xpress GX7 to GX9. There are also satellites in production, especially GEO VHTSs, that have begun to embark elements of a digital payload—for example, Eutelsat's Konnect VHTS, which carries a digital processor. Most embarked digital processors are transparent (i.e., they do the processing down to baseband so that demodulation and modulation are not included). The only known exception is the Starlink, which appears to be regenerative.

This chapter focuses on the full digital payload without addressing hybrid payload architectures. As can be imagined, there are various shades of digital implementation. Section 10.2 examines the definition of the architecture of a digital HTS payload in contrast with that of a conventional HTS payload, overviewing the key elements: the active antenna with the active elements and the digital processor. Many of the elements of a digital payload can be generic, modular, and scalable. Section 10.3 looks into the merit of such an approach. In any satellite, LEO or GEO, the main resource is DC power; Section 10.4 covers RF power efficiency. This book would not be complete without a discussion of the LEO-GEO debate. Accordingly, Section 10.5 delivers this discussion—with the caveat that it is too early to reach a definitive conclusion on this issue. Only time will tell!

10.2   The Digital Payload

The digital payload represents an enormous leap from the technology of conventional payloads. The concept of a flexible digital payload enabling reconfiguration of the frequency plan, routing, and agile beams that can allocate capacity where and when required has been on the drawing board for a while [1] but is now beginning to make tangible headway.

Of course, looking at the signal path of a payload, the main functionalities are present in both cases, but the way these building blocks are implemented is considerably different. Table 10.1 compares the functionalities of the digital and conventional payloads.

Table 10.1
Comparison of the Functionalities of a Digital Payload and a Conventional Payload

Digital Payload	Conventional Payload
Uplink antenna elements	Uplink antenna
LNA system	LNA system
Analog-to-digital conversion	Down-conversion
Uplink beamforming network	
	Demultiplexing
Channel filtering, routing, and switching function	Channel filtering, routing, and switching function
Downlink beamforming network	
Digital-to-analog conversion	
High-power amplification system	High-power amplification system
	Output multiplexing
Downlink antenna elements	Downlink antenna

One of the salient features of a digital payload is the antenna. In general, the conventional payload may operate in dual polarization, and therefore has a dual-port antenna for the uplink or downlink function. The uplink and downlink functions can be combined in a single antenna system so that the antenna may have four ports for dual-polarization uplink and downlink functionality.

In a conventional HTS payload, a SFPB configuration is a frequent choice that indicates that there are as many feeds as the number of beams that the antenna covers. Additionally, there is a compromise between the size of the feed and performance, so that often three or four antennas are required to produce a contiguous lattice of cells. Each antenna produces a lattice of cells that is sparser to allow larger feeds to be used, whereby the set of antennas produces an interlace of lattices that together delivers the contiguous lattice. Four reflectors allow larger feeds than is possible with three and therefore better performance, but the higher number requires more real estate. The frequency color scheme of HTS systems often implies the usage of a single polarization for the uplink and orthogonal polarization for the downlink so that a feed is used for the uplink and the downlink of a given beam. This translates to each feed being a two-port device with orthogonal polarizations for uplink and downlink. If the gateways use the same frequencies, some feeds may also be used for gateways, and these will have four ports, assuming the diplexers are external.

Multifeed per beam (MFPB) can be used. In this case, the required feed illumination per beam is made up of several feeds with the peripheral feeds being shared with adjacent beams. A simple beamforming network is associated with each beam. The issue of mechanical interference for the feeds in the SFPB system is resolved by using several antennas. In MFPB, the issue is resolved through the sharing of the feeds. Usually two antenna systems are used in MFPB, one for the uplink and another for the downlink.

In both cases, the antenna systems are hardwired for a predefined coverage, and the only flexibility that is available is through selecting which beams are active.

The digital payload employs active phased arrays that allow considerably more flexibility through the generation of a multitude of beams from the same aperture. These beams can be identical or different and can be electronically steered to put the capacity where it is required according to the capacity demand profile at the given time.

The active phased array can generate focused beams where high capacity is required and broader beams where the capacity demand is lower while maintaining seamless coverage. This allows for better commercial use of the satellite resources as the high capacity demands in localized areas can be met while the lower capacity requirements over the rest of the coverage can also be satisfied.

Another technique that can be employed—especially when a large service area is required with several communication nodes whose locations are not easy to predict—is beam-hopping [2, 3]. This is a technique where a beam is time-multiplexed between several cells. For a given beam, the dwell time or time slot can be updated according to the capacity profile of the beam-hopping set of cells to maximize the capacity usage. If the capacity of the beam is fixed, this means that the capacity of the beam can be apportioned among the cell set according to the demand. Digital payloads facilitate the beam-hopping process, and the capacity of the beam does not have to be fixed but can be allocated as required. This provides unprecedented flexibility, especially on the ocean where the total area to be covered is extremely large, but the capacity density is relatively low, and the locations of the vessels and aircraft may not be known to the service provider. It is also attractive in governmental and humanitarian applications where the expected service area may also be very large, but the number of theaters and the area of the actual theaters may be small and unknown at the system-design stage.

Note that a conventional payload has two main subsystems: the antenna system and the repeater. The repeater input typically feeds to the LNA input (often through some cover filter) and the repeater output is the output of the output multiplexer (OMUX). This division into two subsystem facilitates testing. In a digital payload, this distinction is less obvious. The LNA is often included with the antenna elements and therefore within the uplink antenna system. Similarly, the HPAs may be included with the associated feeds. Technically, the repeater can be defined as being between the LNA outputs and the HPA inputs. However, there is a multitude of LNA interfaces and similarly a multitude of HPA interfaces, so that the digital payload often adopts a more integrated approach.

Getting to the core of the conventional payload, the down-conversion is followed by a demultiplexer. For a forward payload, this would demultiplex the gateway composite signal to the user beam signals while for the return payload, the individual user beam signals are multiplexed into a composite signal as intended for a given gateway. There is little flexibility available in the filtering, as in the hardware world one filter corresponds to one bandwidth, and thus such flexibility is equipment-intensive and expensive.

In a digital payload, the digital processor is bounded by the analog-to-digital conversion (ADC) at the input and the digital-to-analog conversion (DAC) at the output so that the beamforming network (BFN) function moves from the analogue domain to the digital domain, providing more flexibility. The combined functions of the ADC and the DAC can be equated functionally to the down-conversion in a conventional payload. The channelization and routing functions are performed through digital signal processing (DSP) in the digital domain. Usually an elemental bandwidth is defined to be a few megahertz, and elemental channels can be combined to create the required bandwidth. The channelization, gain control, and routing is performed at this level.

The uplink signal from the gateways and the user beams feed into the processor, which in turn generates all downlink signals for the gateway and the user beams. This means that the segmentation that exists in conventional HTS payloads between the forward and return signal paths does not necessarily exist in a digital payload. Meshed communications between users and even between gateways can, in principle, also be supported.

Figure 10.1 depicts the digital payload, which reduces to three major systems: the uplink antenna system, the digital processor, and the downlink antenna system. Notice that the analog domain of the payload is limited to the antenna sections. The digital domain occupies the core section so that as many functions as possible are performed in the core section, which uses DSP allowing for more flexibility. It is called a digital payload, because the uplink signal is converted to a digital signal as early as possible, and the downlink signal is translated into analog as late as possible, which maximizes the processing in the digital domain.

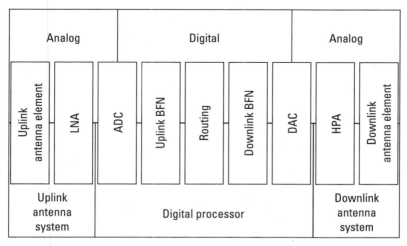

Figure 10.1   A functional diagram of a digital payload.

10.2.1   Active Antennas

Figure 10.2 illustrates the functional building blocks of a receive phased array. In a conventional reflector antenna, each beam is associated to one feed in SBPF (or the MBPF of a few feeds). Referring to Figure 10.2(a), in a phased array, there are many feeds or antenna elements, often tens or hundreds to form a planar array shared by all beams. Each antenna element is followed by two control elements, one to control the amplitude and another to control the phase. The resulting signals from each antenna element are then summed together to constitute the beam signal. The network, including the amplitude and phase control and the summer, is referred to as the BFN.

Of course, the phase and amplitude coefficients determine the shape and the direction of the beam. The same antenna elements can be used to generate several beams by having appropriate amplitude and phase coefficients applied corresponding to the different beams. The advantage here is that the coefficients allow unlimited options for the beams, and although the antenna is optimized for its required field of view, the resulting operational flexibility is considerable.

Phased arrays have used hardware to control the amplitude and phase where the number of beams has been limited by the number of available physical BFN networks [4]. In a digital payload, these functions tend to become DSP functions, allowing even more flexibility and reducing the hardware complexity of the phased array. The complexity is transferred to digital processing, which has its own issues and consumes DC power.

Figure 10.2 depicts two schemes for a receive phased array. The transmit version is the mirror image. Figure 10.2(a) shows a passive phased array, which although offering full flexibility, may limit the performance. Having the BFN prior to the LNA would increase the losses and degrade the G/T performance, while in the transmit version having the HPA prior to the BFN would increase the output losses and the power efficiency of the payload.

The active phased array improves the performance by putting the active element just next to the antenna element as shown in Table 10.1 and illustrated in Figure 10.2(b). The top four entries of the digital payload column of Table 10.1 represent the functions of the uplink active antenna, while the last four entries represent the downlink active antenna.

The difference between the two options is that for the passive phased array, the number of LNAs is equal to the number of receive beams and the number of HPAs is equal to the number of transmit beams, while for the active phased array, the number of LNAs and the number of HPAs is equal to the number of the antenna elements for the receive antenna and for the transmit antenna, respectively.

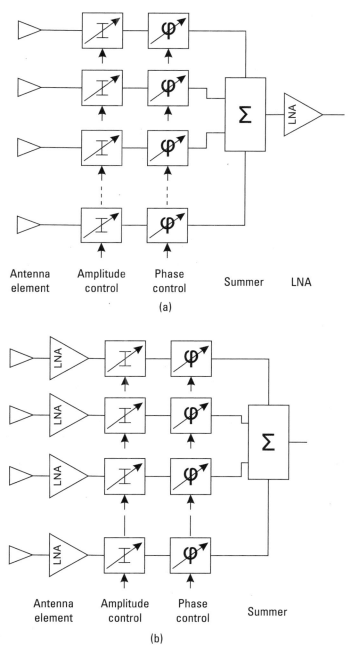

Antenna
element
Amplitude
control
Phase
control
Summer　　LNA

(a)

Antenna
element
Amplitude
control
Phase
control
Summer

(b)

Figure 10.2　The functional building blocks of a receive phased array: (a) passive version and (b) active version.

Although some tolerance exists in the variance of absolute amplitude and absolute phase performance of the RF chain of the elements, calibration techniques and temperature control become an important part of the system to ensure adequate performance through the diurnal temperature variations and longer term aging effects.

Since signal components potentially employ all antenna elements and consequently all RF chains carry contributions from all beams, the elements cover the full frequency plan, and therefore an active antenna is required to be wideband. This presents challenges for each element of the RF chain but simplifies the payload architecture.

Phased arrays can be direct radiating arrays (DRAs) or can constitute the feed system of a reflector antenna to form an array-fed reflector antenna (AFRA). DRAs can offer considerable scan performance, which may be an asset for LEO system. Since the gain of the antenna is a function of the aperture, in theory, a DRA could have the elements it needs to cover the aperture area. However, DRAs may not be practical for large apertures since the number of elements increases to the square of the linear dimension of the array. An AFRA offers a better compromise for large apertures. On a GEO satellite, an AFRA is a good option since higher gain and less scan performance is required.

Figure 10.3 shows an unfurlable mesh antenna with a diameter of 11m that has been produced by Northrop Grumman Astro Aerospace for L-band applications. When stowed, the reflector diameter is less than 1m, and the mass of the reflector and the articulating boom (not shown in Figure 10.3) is

Figure 10.3   An 11-m unfurlable perimeter truss reflector. (*Source:* Northrop Grumman Astro Aerospace. Reprinted with permission.)

less than 100 kg. There are several factors in a reflector design that impact the maximum frequency it can support. Similar technology has been developed for the Ka-band and beyond. With this technology, as the operational frequency of the antenna increases, a higher density mesh is required.

Another range of unfurlable reflectors is available from L3Harris with diameters ranging up to 25m. The design of these mesh reflectors is compatible with Ka-band operation. Figure 10.4 depicts such a reflector.

By nature, as the aperture increases, the antenna pointing becomes increasingly important. This has its own challenges, but phased arrays lead

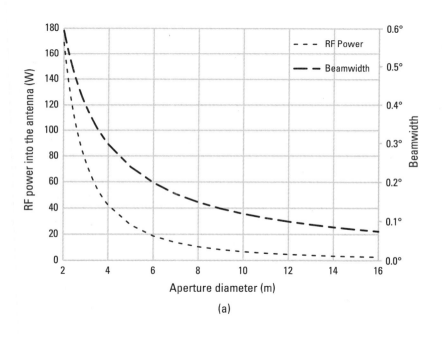

(a)

(b)

Figure 10.4   Rendering of a 5-m Ka band radial rib reflector. (*Source:* L3Harris. Reprinted with permission.)

themselves more easily to active beam pointing. The phased array can be used as an interferometer to detect where the signal is coming from and then use this information for beam-pointing corrections. This, therefore, becomes another function of the processor which electronically steer the beam to the correct director with no moving parts.

Phased arrays cannot only steer beams and shape them, but they can also null interference (or jamming) from different locations. This is, of course, important in military applications where this function was initially exploited, but for HTSs it can be used as spatial interference–mitigation or –management technique.

10.2.2 Active Elements

One of the main considerations for a phased array is the number of beams to be generated. When the number of beams is large (e.g., thousands), the number of antenna elements may be smaller (e.g., hundreds). Thus, from a hardware point of view, the phased array becomes attractive.

Chapter 8's discussion of smaller beams is also valid for GEO. Figure 10.5 displays the required RF power into the antenna and the beamwidth for varying antenna apertures. Table 10.2 lists the assumptions, which are kept similar. For this analysis, we assume that the performance of phase array in terms of gain is similar to a reflector antenna with similar aperture dimensions.

As the aperture dimension increases, the required (aggregate) RF power into the antenna and the beamwidth decreases in an exponential fashion. It will be noticed that the 100-W RF required to maintain the link budget with a 2.6-m antenna aperture reduces significantly as the aperture dimension increases, as can be seen in Table 10.3.

The antenna apertures considered go beyond any known satellite program, but the comparison demonstrates what such an enabling technology

Table 10.2
Link Budget Assumptions

Frequency	19.70 GHz
Antenna efficiency	65%
Payload output losses	1.7 dB
Range	37,236 km
Earth station G/T	16.1 dB/K
C/N_0	106.0 dBHz

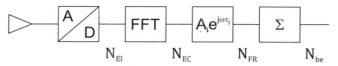

Figure 10.5 RF power into the antenna and beamwidth as a function of antenna aperture.

could deliver. The previous section highlighted the availability of the technology for such apertures. As discussed for the LEO case in Chapter 8, if we were to populate a service area with the smaller beams, then the required RF power would be the same. If we take the GEO case with a 2.6-m antenna aperture as the reference case and consider the 10-m case, using the numbers in Table 10.3, it would require 33.3 beams to cover the corresponding 0.46° beamwidth, i.e. $(15/2.6)^2 = (0.457/0.0792)^2 = 33.3$. The 10-m case requires 3-W RF for this reference link budget, and this power for the number of beams yields a 100-W RF (i.e., 3×33.3–W RF), which is the power required for the reference case. In a broadcast case, it may be simpler to stick to the smaller antenna that corresponds to the service area. In an HTS, the information is probably only intended to a single user, so having a large aperture allows the use of minimal resources to deliver the content and an increase in capacity and the number of users served. In the above example, if the same bandwidth is available per beam in both cases, the system capacity is increased 33.3 times, and the number of users can be potentially increased by the same factor. This may not be realistic as more spectrum may not be available. If the same system bandwidth is assumed in both cases, and the smaller beam lattice uses a four-color scheme frequency plan, the system capacity increases by 8.3 times, and the number of users follows suit. The enhanced performance is still significant.

Narrow beams are thus attractive for efficient communications where high capacity is required. Broader beams can be used to ensure that the coverage is seamless where the capacity demands are lower. At the antenna level in the GEO case, this means that a large reflector is attractive together with a large feed array. This allows better commercial use of the satellite resources as

Table 10.3
The Required RF Power into the Antenna to Maintain the Peak EIRP for a Range of Aperture Diameters and the Corresponding Beamwidths

Antenna diameter	2.6m	5m	10m	15m
Beamwidth (4 dB)	0.46°	0.24°	0.12°	0.08°
RF HPA power	100W	27W	6.7W	3.0W

the high capacity demands in localized areas can be met while the coverage where low capacity is required can also be satisfied, although it comes at a cost!

In conventional payloads, an active element has an operating point that is a function of the loading of the respective beam for the required linearity. However, in a digital payload, each active element has an operating point that is a function of the system loading. The fact that the linearity of the system is a function of the system loading is pertinent to both the HPAs of the downlink active antenna and the LNAs of the uplink active antenna. It represents a better system trade-off.

The spacecraft TWTA can deliver more than 100-W RF with an efficiency of more than 60% at the Ka-band and has been the workhorse of communication satellites. Boeing has accumulated 80.5 million on-orbit hours with 1,783 TWTAs over a period of 20 years [5], and therefore the data available is extensive. TWTAs have proved to be reliable HPAs.

The RF power of SSPAs is typically an order of magnitude lower than that of TWTAs but tends to be smaller and more linear and is often considered to be better suited for active antennas.

Thus, referring to Table 10.3, the left entries of the table would typically be TWTAs while SSPAs would become increasingly attractive moving to the right.

If we consider the forward section and assume the amplitude profile across the aperture, it is likely that the operating point of the edge elements is lower than that of the central elements if the same HPA type is used throughout. This means that the linearity of the edge elements is higher than that of the central elements, if all active element are rated at the same power level. For a reduction in the operating point of 1 dB, the nonlinear products decrease by 3 dB, and consequently the C/I increases by 2 dB. As the nonlinear contributions decrease faster than the wanted signal with a decreasing operating points, the aperture illumination of the intermodulation products is different than that of the wanted signal and will have a lower edge taper in the corresponding illumination profile. The result is that the beam associated with the nonlinear emissions is broader than that of the wanted signal with an ensuing improvement in the C/I.

SSPAs are attractive for active antennas because of their ease of integration in a transmit unit. There is an increase in interest in wide energy gap devices like gallium nitrite (GaN), because they offer better efficiencies at the frequencies of operation. Gallium arsenide (GaAs) has been used for a longer time. This presents its own issues for space applications since there is less experience with GaN. Table 10.4 provides a comparison of the salient parameters of both materials.

Table 10.4
Comparison of Key Parameters for GaAs and GaN [6]

Parameter	GaAs	GaN
Energy gap	1.4 eV	3.4 eV
Breakdown electric field	0.5 MV/cm	4.0 MV/cm
Saturation velocity	2.0×10^7 cm/s	2.7×10^7 cm/s
Thermal conductivity	0.5 W/cm/K	1.7 W/cm/K

The RF powers available from SSPAs are lower than the power from TWTAs, but in an active antenna, the required RF power from each HPA decreases with the number of elements. GaN devices can deliver RF power using simpler topologies with less transistors and therefore less losses and thermal issues [5]; accordingly, these devices are gaining ground for applications of tens of watts at the Ka-band. Below 1-W RF GaAs offer a more cost-effective solution. Below 100 mW, silicon germanium becomes attractive, particularly for LEO systems.

The energy gap is a metric of the power density and radiation hardness, while the breakdown electric field and the saturation velocity point toward higher operating voltages and current density, respectively. Having the right material is a good start but not the total solution. Higher operating voltages not only improve HPA efficiency but also improve the efficiency of the DC power–conditioning system. The heat transfer to the substrate is essential to ensure that the smallest chip dimensions with the highest power and better efficiency can be extracted. Diamond offers better thermal conductivity with respect to the currently used silicon, silicon carbide, sapphire or GaN, but is more expensive [8]. The die bond materials can also improve the thermal conductivity. These parameters show that GaN is superior for higher power, higher frequency, and small dimensions at higher power efficiencies in the space environment.

On the return link, the uplink power from a given beam is shared among the elements of the array so that if the number of elements exceeds the number of beams, the operating point of a given LNA device would be lower, and therefore the linearity would be better. Thus, the noise figure becomes a function of the device and the linearity of the device, as usual, but also a function of the number of elements (and the device). This adds another dimension to the optimization of the uplink antenna.

When considering large active antennas, the classical redundancy approach becomes impractical. Since the system is intrinsically distributed, there are no items that are on the critical path. A failure of one transmit element can be compensated for by the other element, perhaps with a reoptimization of the antenna coefficients. This leads to a more realistic approach of graceful degradation. The system is designed to deliver 100% of the capacity and will deliver a lower percentage (e.g., 90%), assuming failures over the satellite lifetime. That is a pragmatic approach. The more traditional and perhaps the purist approach may be that the system is designed to deliver 100% of the capacity assuming the failures at the EOL. However, unlike conventional payload, there is no redundant unusable capacity in a digital payload, and therefore it makes commercial sense to exploit fully the total capacity that is available at any stage in the satellite lifetime.

Because each element sees a multitude of signal components, the resulting envelope of the combined signal can exhibit a high peak-to-average power ratio (PAPR). Thus, the operating point of the amplifier becomes a trade-off between linearity and the power added efficiency (PAE) for the signal structure considered. An amplifier configuration that is becoming popular in cellular systems is the Doherty topology [9]. Briefly this includes two amplifiers, an input network, and an output network. One amplifier is operated in class AB or B and optimized to work up to a certain level, and then it clips or saturates, while the other operates in class C to complement the signal where the other saturates. The two amplifiers are termed the carrier amplifier and peaking amplifier, respectively. This amplifier configuration allows for increased power output with a higher PAE over a wide range of input signals and is therefore well-suited for active antenna applications.

10.2.3 Digital Processing

The function of the digital processor is to perform at least the following functions:

- Uplink BFN function, including beam pointing corrections;
- Channelization;
- Gain control including fixed gain control and automatic gain control;
- Routing;
- Downlink BFN function, including beam-pointing corrections.

Figure 10.1 shows direct connections between the LNA and the ADC, and between the DAC and HPA. This is called direct sampling or RF sampling, and at least theoretically, it offers the simplest architecture with the highest flexibility. However, the sampling rate of the ADC and the DAC must be compatible with the associated RF signals. Assuming Ka-band, the required sampling rate for the ADC would be more than 60 Gsps (giga samples per second) and for the DAC in excess of 40 Gsps to respect the Nyquist sampling criterion. Some suppliers propose direct sampling for the DAC, but the ADC sampling rate remains challenging. The situation becomes even more challenging when we consider higher bands for the gateways, V-band and/or W-band where sampling rates of 200 Gsps would be required.

The alternative is to use RF converters so that the frequency band is brought down to a band that is compatible with the ADC or the DAC, as the case may be. At the V-band and W-band, this is expected to be the preferred solution. Direct sampling employs the latest devices, and therefore the technical and industrial choice may be limited. Using converters may open the door to more established devices with a bigger choice in the number of bits and therefore enhance the DSP performance. The power consumption of ADCs increases with the sampling rate so there is also a power trade-off between high–sampling rate ADCs and lower–sampling rate ADCs with converters.

The number of ADCs and DACs is often equal to the number of elements of the uplink and downlink antenna, respectively. In some cases, the spectrum from two or more beams can be stacked together (or frequency-multiplexed) to form a single signal to the ADC and the reverse process for the DAC to drive several beams. This technique is compatible with the converter technique, although it tends to be limited to narrower-band applications and impractical with direct sampling.

Referring to Figure 10.2(b), the combined functions of phase and amplitude control translate into a complex multiplication in the digital domain. Figure 10.6 illustrates a simple functional block diagram for a digital BFN system. The input originates from the antenna system with N_{EI} elements, and therefore there are N_{EI} ports to the digital processor with an ADC at each port. The basic spectral processing unit of a digital processor is the elemental channel, so we assume that is performed with a fast Fourier transform (FFT) to produce N_{EC} elemental channels. Thus, the elemental channel defines the granularity of the spectral processing and the smallest bandwidth of a

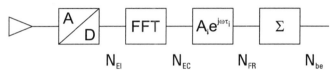

Figure 10.6 Simple block diagram of a digital BFN.

signal to be processed. If there is a frequency reuse of N_{FR}, each signal from an elemental channel is used N_{FR} times to generate N_{FR} components of N_{FR} beams at that frequency.

Complex multiplications are costly computational functions so that the number of complex multiplications in a given time can be used as a metric to dimension the BFN requirements for the DSP. Thus, assuming dual-polar operation, the number of complex multiplications that are required would be

$$M_{BFN} = 2N_{El}N_{EC}N_{FR} \qquad (10.1)$$

The sampling rate of F_{Sam} sps must meet the Nyquist sampling criterion so that if the bandwidth at each ADC is F_{BW}, then

$$F_{Sam} = \gamma F_{BW} \qquad (10.2)$$

where γ is the Nyquist factor and must be greater than two.

The capability of the digital processor for a BFN function (uplink or downlink) would thus need to have the capability of performing M_{BFN} in a sampling interval, δt, or

$$\frac{M_{BFN}}{\delta t} = 2N_{El}N_{EC}N_{FR}F_{Sam} \qquad (10.3)$$

This simple example shows that the computational capability of the processor for the BFN function is independent of the number of beams and ultimately only dependent on the total bandwidth to be handled by the system (i.e., $2N_{El}N_{EC}N_{FR}$). The variables N_{El} and N_{FR} are normally fixed by the service area requirement so that N_{EC} becomes a dimensioning parameter and is typically limited to a few megahertz.

Digital processors can be used for spectral interference mitigation since elemental channels on a specific beam at a specific frequency or a band of frequencies can be blanked out or reduced in gain to minimize the interference degradation effect.

Recall that the functionality and the flexibility of the payload are greatly enhanced through the digital processor. However, as often in life, nothing comes for nothing. The digital processor for an HTS often consumes a few kilowatts. Over the years, the figure has not changed very much, but the capacity handled by such units has increased significantly. In real terms, the DC power required per unit capacity has decreased as the processor technology has improved, and this trend is likely to continue.

10.3 Generic Elements

Satellite operators experienced a reduction in capacity revenues of 18% in 2018. In addition, the commercial satellite market shrunk by 8% and the launch market by 7% over the period 2012–2016 [10]. In the past, 20–25 GEO satellites were procured each year, but only 14 satellites were procured over the two-year period of 2017 and 2018; 2019 has been better, with 15 satellites procured [11]. All this shows the commercial pressure on satellite operators to rethink their business strategies and on the space industry to innovate to meet the evolving technical, commercial, and economic requirements.

Satellite production has been very bespoke with procurement processes that are specific to a single satellite with its own operational requirements and procurement requirements. The frequency plan, the service area, and the associated performance are often specific to one satellite or at most a couple of flight models. The first six months of a satellite program typically lead to the baseline design review (BDR). As the name implies, this is the design of the specificities of the satellite because each satellite is different. This period typically translates to 20% of the schedule with the associated costs. Obviously, this method of working comes with expense. The cost of capacity acquisition can be decreased through genericity. LEO systems with hundreds or even thousands of identical satellites are more amenable to such production techniques.

Looking at Figure 10.1 for a digital GEO HTS, the antenna elements together with their associated active elements are wideband and therefore could cover satellites intended for any given band (i.e., the Ka-band). The numbers involved on given satellites are already large, and creating generic compact integrated design could become the baseline for Ka-Band HTS satellites. We could go a step further and produce a standard tile or subarray. Customization could be through the aperture size of the array.

The digital processor is characterized by the capacity it can handle and the number of input and output ports, which translates to the number of ADCs and DACs, respectively. It is intrinsically modular. The number of input and output ports can be modular, and the processing section is scalable. Of course, a given design has limits in the maximum capacity and the number of ports that it can handle.

The digital payload leads itself to a generic payload where a large number of elements could be standardized. There are a number of space industry initiatives in this direction (e.g., OneSat from Airbus [12] and INSPIRE from Thales Alenia Space [13]). mPOWER from Boeing also has elements that could make it generic.

In conventional HTSs, the user service area could be comprised to be usable from several orbital locations, but mapping the gateway locations from different orbital locations is particularly challenging and often either practically impossible or extremely constraining. Managing capacity across user beams is also difficult and limited, because it becomes hardware-intensive, but this is part of the functionality offered by the digital payload. In fact, the flexibility of a digital payload is such that traffic can be concentrated to limit the number of gateways in the ramp-up period and then spread to more gateways as traffic builds up.

10.4   Payload RF Power Efficiency

The main resource on a spacecraft is DC power, and therefore its efficient use maps to more capacity and more scope for revenue. The efficiency of the HPA has been under great scrutiny. The high gain of the TWTA makes the input power negligible with respect to the output so that the efficiency can be expressed as the ratio of the RF power to the DC power.

The gain of a SSPA is typically 10 dB or less so the input power is no longer negligible, and it is more customary to speak of PAE, where the input RF power is deducted from the output RF power.

The HPA contributes to only one component in the useful RF power equation, which can be expressed (in decibels) for the nth antenna element of an active antenna as

$$P_{\text{RFU}n} = P_{\text{HPAR}n} + L_{\text{OP}n} + OBO_n + L_{\text{Ant}n} \tag{10.4}$$

where:

$P_{\text{HPAR}n}$ is the rated RF power of the nth HPA;

$L_{\text{OP}n}$ represents the output losses between the HPA output and the antenna element in the nth chain;

OBO_n is the output back-off or the reduction with respect to the rated RF power of the nth HPA;

$L_{\text{Ant}n}$ includes all the losses of the nth antenna element;

OBO_n includes two contributions: one that may be considered constant for all elements to determine the HPA system linearity and another that is a function of the location of antenna element to determine the aperture illumination profile.

Note that the parameters $L_{\text{OP}n}$, OBO_n, and $L_{\text{Ant}n}$ have negative values in decibels.

In a phased array we need to consider the total useful power into all the antenna elements so that (10.4) becomes

$$P_{\text{RFUT}} = \sum_{n=1}^{N} P_{\text{HPAR}n} l_{\text{OP}n} obo_n l_{\text{Ant}n} \tag{10.5}$$

where N is the number of elements in the array.

Note that lowercase letters are used to denote linear values and that $l_{\text{OP}n} < 1$, $obo_n < 1$, and $l_{\text{Ant}n} < 1$.

The numbers in Table 10.3 for the EIRP may be considered to be as given in the above equation.

Looking at the total HPA system, the power conversion efficiency can be written as

$$\eta_{\text{HPASys}T} = \frac{\sum_{n=1}^{N} P_{\text{HPAR}n} l_{\text{OP}n} obo_n l_{\text{Ant}n}}{\sum_{n=1}^{N} P_{\text{DC}n}} \tag{10.6}$$

where $p_{\text{DC}n}$ is the DC power consumption of the nth HPA.

The thermal dissipation of the HPA system is given by

$$P_{\text{Therm}T} = \sum_{n=1}^{N} P_{\text{DC}n} - \sum_{n=1}^{N} P_{\text{HPAR}n} l_{\text{OP}n} obo_n l_{\text{Ant}n} \tag{10.7}$$

From a system point of view, the objective would be to maximize (10.6), which, in turn, would minimize (10.7). The transmit *module* principle, which puts the HPA in proximity to the antenna element, helps reduce the post HPA losses. When possible, phase aperture profiling is preferred to operate the HPAs as close as possible to the rated power (i.e., to maximize all values of obo_n). An alternative is to optimize so that the sum of all beam amplitude contributions to each antenna element is maintained as close as possible to the optimal operating point of the HPAs.

Table 10.5 compares HTS payloads employing the conventional and digital approaches. An integrated wideband transmit design approach for the digital payload makes the RF chain more compact, and reduces the losses, and therefore contributes positively to the power conversion efficiency. However, the concentration of the transmit elements of the phased array in a relatively small area, and consequently a compact volume with the ensuing dissipation, entails considerable thermal design effort.

Both conventional and digital approaches require HPAs to operate in multicarrier operation for both the forward and the return paths. In a

Table 10.5
Comparison of a Conventional HTS Payload and a Digital HTS Payload

Parameter	Conventional HTS Payload	Digital HTS Payload
L_{OPn}	High	Low
Associated performance degradation due to flexibility	High	Low
Associated performance degradation due to multicarrier operation	Low	High
Spacecraft antenna pointing	At reflector level	At beam level
Service area	Predefined	Flexible
Channelization	Predefined	Flexible
Routing	Predefined	Flexible
Managing dynamic capacity demands	Difficult	Available
Interference mitigation	Difficult	Available

conventional payload, the linearity requirements for the return paths are higher than that for the forward, because the return signal multiplex is larger. Typical output back-off rates for TWTAs are under 2 dB and about 4.5 dB for the forward and return cases, respectively. In a digital payload, the linearity requirements are more akin to that of the return in the conventional payload. As SSPAs tend to be more linear than TWTAs, the gap narrows.

For a digital payload, although the service area for a GEO system could be the visible Earth as seen by the satellite, it is often optimized over a smaller area. Managing dynamic capacity demands and interference mitigation with conventional payload tends to be equipment-intensive with a mass penalty and is often limited to the bare essential. The digital payload is more amenable to these functions. It also can offer interference mitigation at two levels:

- Using the active antenna to spatially null the interference source or sources;
- Filtering out spectrally the interference.

10.5 The LEO-GEO Debate

There are three LEO systems in the making, namely Starlink from SpaceX, Amazon's Kuiper, and Telesat LEO (obviously from Telesat). A fourth,

OneWeb has gone through the Chapter 11 bankruptcy process but recently been acquired by Her Majesty's Government, through the U.K. Secretary of State for Business, Energy and Industrial Strategy and Bharti Global Limited. There are also a number of major GEO HTSs in production including Eutelsat's Konnect VHTS, three ViaSat 3 satellites, and Jupiter 3 from Hughes. All this capacity will be available in the coming years.

Konnect VHTS, ViaSat 3 satellites, and Jupiter 3 will each provide between 0.5 and 1 Tbps. The COVID 19 pandemic has had an impact on the schedule of these satellites, but it is expected that each will have a launch in 2021 and probably be operational by 2022.

Telesat LEO has announced between 16 and 26 Tpbs of system capacity [14]. SpaceX in its application to the Federal Communications Commission (FCC)[1] [15] put a capacity objective per satellite of between 17 and 23 Gbps, with 20 Gbps being an average. If we assume the current production rate of Starlink satellites as the determining factor, this is currently at 120 satellites a month [16]. At the current launch rate of 60 satellites at a time, this requires two launches a month. Since both Starlink and SpaceX belong to Elon Musk, it can be assumed that this can be managed! Thus, with the current industrial capacity, there will be 1,440 satellites a year launched, and assuming that the ground infrastructure is not on the critical path, at this rate, there would be 28.8 Tbps of system capacity a year.

However, the industrial effort to implement such a system represents an extraordinary feat. Figure 10.7 shows a hypothetical demonstration based on the Starlink model. At the announced 120 satellites a month production rate, it would take 8.3 years to manufacture 12,000 satellites, but with an assumed lifetime of five years; after five years, the production rate would have to be doubled to maintain the deployment phase of 8.3 years for the full constellation. With this assumption, the first satellites would reach the end of their life before the constellation became full! Thereafter, the replenishment process is continuous to maintain the system. A current GEO satellite has a design lifetime of 15 years, although it may be used for longer. Under these assumptions, the LEO constellation would require 53,760 satellites to be launched for a full operational phase of 15 years and nearly 900 launches at 60 satellites a launch. An average production rate of 2,400 satellites and 40 launches a year would be required to maintain the constellation.

---

[1] The FCC is an independent US agency that regulates interstate and international communications by radio, television, wire, satellite and cable. It is also the notifying administration to the International Telecommunication Union (ITU) that coordinates the utilization of spectral and orbital resources at the international level.

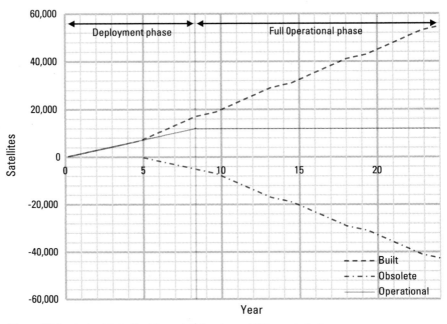

Figure 10.7   Industrial effort to maintain a large LEO constellation.

This is an example based on the information available and is not necessarily representative of the Starlink case. Nevertheless, it demonstrates the industrial and associated financial challenges of maintaining such a constellation.

The capacity of LEO systems, especially inclined orbit constellations, is evenly spread on all longitudes and offers symmetrical coverage about the equator, possibly extending to the pole to make it global. This is an important feature since GEO systems cannot offer such services. There are approximately 4 million people [17] living in the Artic; that is, above the Artic Circle at a latitude of 66.34°N, and this corresponds approximately to the limit of GEO coverage (with elevations of 15° or less).

There are commercial shipping interests in the Artic. The Northeast Passage is a sea passage from the North Atlantic to the Pacific along the Norwegian and Russian coast that serves as an alternative route to the Suez Canal. Part of the Northeast Passage overlaps with the Northern Sea Passage, which is the Russian Exclusive Economic Zone. Considering routes from Rotterdam, the Netherlands to Shanghai, China, Yokohama, Japan and Busan, South Korea, the Northeast Passage is about 30% shorter than the route through Suez Canal. The Northwest Passage is another Artic route between the Atlantic and Pacific oceans along the northern coast of North America and an alternative to the

Panama Canal. Although meteorological conditions may introduce uncertainties in the availability and in the maritime duration of the passage, climate change is expected to make these passages more navigable.

Flights connecting cities in Asia and North America also use north polar routes.

There are only a few thousand people in the Antarctic Region, and commercially there is less activity.

The polar market is an ideal target market for global LEOs and practically virgin. It is probably the reason for the polar orbit satellite in the Telesat LEO system. In fact, 40% of Canadian territory with more than 200,000 inhabitants [18] is in the Artic Region.

By nature, LEO constellations, especially global systems, are well-suited for governmental, first-responder, and humanitarian applications with a single terminal that can be deployed practically anywhere in the world and be rapidly operational.

The Earth is composed of 71% of ocean and 29% of land [19]. Additionally, 95% of the world's population lives in 10% of the land [20]. The Earth's population is currently estimated as 7.8 billion, and the Earth's surface is 510 billion km^2. This means that 7.4 billion people live on less than 3% of the Earth's surface or 14 billion km^2. This is equivalent to less than twice the area of a European or a U.S. coverage. Of course, the 3% of the Earth's surface, where the large agglomerations, attracts terrestrial infrastructure where services are profitable, but as the population density decreases terrestrial communications become more inefficient and less profitable leaving these areas that are either underserved or completely unserved. Demographics with lower population densities over a large area favour GEO VHTSs as they can be better adapted to meet the specific regional requirements covering the population over a relatively large area. These systems can provide large capacities in a shorter deployment time. Typically, VHTSs can deliver the full capacity in three-and-a-half to four years and provide a service that exceeds 15 years. This makes them more flexible and agile to meet demand and financially more manageable and less risky.

Of course, this capacity can also be delivered by LEO constellations. The challenge is the optimization or dimensioning of such systems. The capacity requirements over the populated areas is considerable higher than the rest of the world. At a simplistic level, a global system that is dimensioned for 3% of the service area is over-dimensioned for 97% of the service area or a system that is dimensioned for 97% of the service area is under-dimensioned for the lucrative 3%. Although techniques exist to concentrate capacity, perhaps even from a cluster of satellites within the constellation over a given area, the challenges are considerable.

It is interesting to compare the capacity densities of GEO VHTSs and LEO constellations. The area of a CONUS or a European service area is about 8 billion or 10 billion km^2. If we consider a GEO VHTS system with 1 Tbps of capacity, the average capacity density is between 100 and 120 kbps/km^2 depending on the case considered. If we now consider a global LEO system, the capacity density is about 20 kbps/km^2 per 10 Tbps of system capacity. Thus, if we take Telesat LEO, the capacity density would be about 40 kbps/km^2, and if we take the Starlink system, after the first year of satellite deployment, the capacity density would be about 56 kbps/km^2 reaching 470 kbps/km^2 with a constellation of 12,000 satellites. So technically a LEO constellation could provide more capacity density than a GEO VHTS system but not efficiently, as discussed earlier. Additionally, certain LEO systems may have the potential to manage their capacity so that where the overall demand is low, a LEO satellite has the potential to allocate the satellite capacity to a few nodes within the service area. This could be an attractive feature for 5G applications in certain areas and for cruise ships, along with aircraft IFC.

As the name implies, the low altitude of LEO satellites contributes to lower latency. The round-trip delay of a 550-km LEO is less than 4 ms while that of a GEO is about 250 ms, but this is only part of the story. In optic fiber the refractive index is about 1.47, which means that the speed of light is slower by the same factor. Table 10.6 gives the delta in the propagation time when using free space and an optic fiber between two cities.

It is interesting to note that the distance given is the shortest distance on a great circle, which would be close to a LEO system perhaps using optical intersatellite links and possibly using satellites as relays. An optic fiber is more likely to follow coasts and geographies, and therefore the actual gap is probably wider than that given in Table 10.6.

We can now compare the propagation times of both systems. For the optic fiber system, we assume the great circle distance between two cities at the Earth's surface. For the satllite constellation system, we assume a great circle distance at the satllite altitude plus twice the altitude of the satellites. The breakpoint in propagation time between the two systems can be given as [21]

$$t_p = \frac{2A_S + 2\pi q\left(R_E + A_S\right)}{c} = \frac{2n\pi q R_E}{c} \tag{10.8}$$

where:

c is the speed of light in free space;

q is the distance between the two sites expressed as a fraction of the Earth circumference;

Table 10.6
Propagation Time Difference Between Free Space and Optic Fiber Propagation
Assuming the Shortest Distance Between the Two Cities in Both Cases

City 1	City 2	Distance	Propagation Time Delta between Free Space and Optic Fiber
Paris	Moscow	2,487 km	4 ms
Paris	New York	5,790 km	9 ms
Paris	San Francisco	8,956 km	14 ms
Paris	Tokyo	9,715 km	15 ms
Paris	Singapore	10,740 km	17 ms
Paris	Sidney	16,966 km	27 ms
New York	Moscow	7,373 km	12 ms
New York	San Francisco	3,965 km	6 ms
New York	Tokyo	10,569 km	17 ms
New York	Singapore	15,095 km	24 ms
New York	Sidney	15,883 km	25 ms

n is the refractive index of the material used for the optic fiber.

Using (10.8), we can now express the satellite altitude as

$$A_S = \frac{(n-1)R_E}{1 + \dfrac{1}{\pi q}} \tag{10.9}$$

Figure 10.8 illustrates (10.9). It should be evident that over long distances, satellites should be faster. Thus, for distances between the two sites to the right of the curve, the constellation would offer better latency, while for distances to the left of the curve a direct optical fiber link would be advantageous. Figure 10.8 also shows the interest in LEO systems especially at lower altitudes. Thus, for example, at a satellite altitude of 500 km, for distances longer than 2,500 km, the satellite system would have a shorter propagation delay and therefore offer better latency. Referring to Table 10.6, this means that practically for all trans-Atlantic, trans-Pacific and transcontinental communications, such a satellite system would offer better latency.

Table 10.7 provides a system-level comparison between LEO constellations and GEO VHTSs.

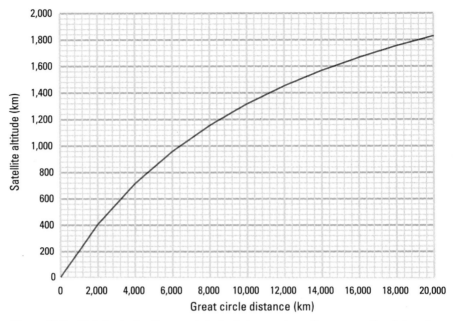

Figure 10.8 Plot displaying the great circle distance where the propagation delays for both a satellite system and an optic fiber are equal.

Table 10.7
Top-Level Comparison Between LEO Constellations and GEO VHTSs

Parameter	LEO Constellations	GEO VHTSs
Coverage for polar regions	Intrinsic for global systems	No
Oceanic coverage	Intrinsic	Possible
High-density coverage	Possible but dimensioning	More agile and rapid deployment
Coverage sweet spot	Large to global	Regional
Capacity density	20 kbps/km² per 10 Tbps	100 kbps/km² per 1 Tbps for continental coverage
Round-trip delay	4 ms	250 ms
Latency sweet spot with respect to fiber	See Figure 10.8. (e.g., at a satellite altitude of 500 km a distance >2,500 km)	250 ms per satellite hop
System cost	10× to 100× M€	× M€
Industrial scale	Larger	Smaller
Deployment time scale	Longer	Shorter

Table 10.7 shows that the HTS landscape will become richer in solutions. LEO systems have the advantage of latency and coverage, even global coverage. However, the deployment phase is long and requires huge capital. GEO systems are relatively more agile to the capacity demand, and a 1-Tbps GEO regional system can deliver roughly the capacity density of a 50-Tbps LEO constellation.

This means that in the (near) future wherever one is on Earth, there will be a choice of HTS systems available. In urban areas, terrestrial communications, especially with the advent of 5G, will become increasing prolific with better speeds and QoS. At some stage, as one leaves the cities, we enter the underserved zone, and further away we get to the unserved region. This is not because terrestrial systems are deficient but because of economics. This is where HTSs thrive. The complementarity of both systems will offer the most cost-effective solution for the given location. Will it be GEO or LEO? There will be a choice. The ultimate solution is a system of systems, terrestrial, GEO, and LEO so that no one is left unconnected wherever one may be.

References

[1]   Fenech, H., R. Hitchcock, and E. Lance, "The Dream Payload," *European Space Component Conference, ESA (ESTEC)*, Noordwijk, the Netherlands, March 15–17, 2011.

[2]   Fenech, H., et al., "VHTS systems: Requirements and Evolution," *11th European Conference on Antennas and Propagation (EUCAP)*, Paris, 2017, pp. 2409–2412, doi: 10.23919/EuCAP.2017.7928175.

[3]   Fenech, H., S. Amos, and T. Waterfield, "The Role of Array Antennas in Commercial Telecommunication Satellites," *10th European Conference on Antennas and Propagation (EuCAP)*, Davos, Switzerland, 2016, pp. 1–4, doi: 10.1109/EuCAP.2016.7481857.

[4]   Montesano, A., et al., "ELSA+: An Enabling Technology for the Flexibility and SW Defined Mission," *38th ESA Antenna Workshop on Innovative Antenna Systems and Technologies for Future Space Missions, ESA/ESTEC*, Noordwijk, The Netherlands October, 3–6, 2017.

[5]   *TWT/TWTA Handbook*, L3 Communications, Electron Technologies, Inc., 2007, https://www2.l3t.com/edd/pdfs/datasheets/TWT-TWTA%20Handbook.pdf.

[6]   Fenech, H., et al., "Satellite Antennas and Digital Payloads for Future Communication Satellites," Special Issue on Recent Advances on Satellite Antennas for Communication, Navigation, and Scientific Mission Payloads, *IEEE Antennas and Propagation Magazine*, Volume 61, Number 5, October 2019, pp. 20–28.

[7]   Browne, J., "What Is the Difference between GaN and GaAs?," *Microwave and RF*, July 7, 2016.

[8] Blevins, J. D., and G. D. Via, "Prospects for Gallium Nitride-on-Diamond Transistors," *2016 IEEE Compound Semiconductor Integrated Circuit Symposium (CSICS 2016)*, Austin, Texas, 2016, pp. 1–4, doi: 10.1109/CSICS.2016.7751059, October 23--26, 2016.

[9] Slade, B., *The Basics of the Doherty Amplifier*, White Paper, Orban Microwave, Inc., https://orbanmicrowave.com/the-basics-of-power-amplifiers-part-3/.

[10] Price Waterhouse Cooper, *Main Trends and Challenges in the Space Sector*, June 2019, https://www.pwc.fr/fr/assets/files/pdf/2019/06/fr-pwc-main-trends-and-challenges-in -the-space-sector.pdf.

[11] Henry, C., "Geostationary Satellite Orders Bouncing Back," *Space News*, February 21, 2020, https://spacenews.com/geostationary-satellite-orders-bouncing-back/.

[12] Airbus website, OneSat, https://www.airbus.com/space/telecommunications-satellites /onesat.html.

[13] Thales Alenia Space website, "Thales Alenia Space Releases Fully Digital Satellite to Address Fast Moving Market Needs," October 9, 2019, https://www.thalesgroup.com /en/worldwide/space/press-release/thales-alenia-space-releases-fully-digital-satellite -address-fast.

[14] Henry, C., "Telesat Preparing for Mid-2020 Constellation Manufacturer Selection," *Space News*, May 1, 2020, https://spacenews.com/telesat-preparing-for-mid-2020 -constellation-manufacturer-selection/.

[15] Application of Space Exploration Holdings, LLC, For Approval for Orbital Deployment and Operating Authority for the SpaceX NGSO Satellite System, *Application for Approval for Orbital Deployment and Operating Authority for the SpaceX NGSO Satellite System*, FCC, November 15, 2016, https://fcc.report/IBFS/SAT-LOA-20161115-00118 /1158349.pdf.

[16] Sheetz, M., "SpaceX Is Manufacturing 120 Starlink Internet Satellites per Month," CNBC Website, 10 August 2020, https://www.cnbc.com/2020/08/10/spacex-starlink -satellte-production-now-120-per-month.html.

[17] National Snow and Ice Data Center, *Artic People, All About Arctic Climatology and Meteorology*, https://nsidc.org/cryosphere/arctic-meteorology/arctic-people.html.

[18] Government of Canada website, *Canada and the Circumpolar Arctic*, https://www .international.gc.ca/world-monde/international_relations-relations_internationales /arctic-arctique/index.aspx?lang=eng.

[19] *7 Continents*, https://www.worldometers.info/geography/7-continents/.

[20] Joint Research Centre news release, "Urbanisation: New World Bank and European Commission Map Shows that 95% of the World's Population Lives on 10% of the Land," European Commission, December 17, 2008, https://ec.europa.eu/jrc/sites/jrcsh /files/jrc_081217_newsrelease_travel_times_en.pdf.

[21] Khan, F., *Mobile Internet from the Heavens*, Samsung Electronics Richardson, Texas, arXiv:1508.02383 [cs.NI], August 9, 2015, https://arxiv.org/abs/1508.02383.

About the Author

Hector Fenech was born in Malta and received a B.S. in engineering from the University of Malta in 1978. In 1981, he was awarded a scholarship at the Phillips International Institute of Technological Studies, Eindhoven, the Netherlands, from which he earned an M.S. in electronic engineering in 1983. He read mobile satellite communications at the University of Bradford, United Kingdom, where he earned a Ph.D. degree in 1987.

In 1989, Fenech joined Eutelsat where his last position was director of Future Satellite Systems. He was responsible for system analysis, mission definition, and negotiations with industry regarding the communication mission requirements of over 35 satellites ordered by Eutelsat including the S-band payload on W2A, the Ka-band HTS mission on KA-SAT, the first of the Ka-band generation, and Eutelsat Quantum, the first commercial software–defined Ku-band mission. In his position, he was responsible for understanding the commercial and economic objectives of the company and worked with industry to determine the most cost-effective solutions. This involved numerous technology discussions with the space industry on both sides of the Atlantic.

Fenech started with Eutelsat as communication systems engineer and subsequently served as a payload manager, monitoring the manufacture of nine satellites. This means that he has seen satellites from conception to operations. Moreover, he was involved with several projects with the ESA and other space agencies.

Before Eutelsat, Fenech headed the Satellite Communication Systems Group at Ferranti International Signal plc., Ilkley, West Yorkshire, United Kingdom, where he worked for three years. From 1983 to 1989, he worked at the University of Bradford as a research assistant and later a lecturer.

Over his professional career, Fenech has been involved in various facets of satellite communications, and he has more than 90 publications and 14 patents. In addition, he is a fellow of the American Institute of Aeronautics and Astronautics, the Institute of Electrical and Electronic Engineers, and the Institution of Engineering and Technology.

In 2016, Fenech was a first-year recipient of the Outstanding Alumni Achievement Award of the University of Malta where he is an affiliate professor in the Faculty of Information & Communication Technology. He has given courses in satellite communications, co-organized the SatNex Summer School of 2017 in collaboration with the ESA, and is working on promoting satellite communications.

Index

Artech House
Space Technology and Applications Series

Space Microelectronics, Volume 2, Integrated Circuit Design for Space Applications, Anatoly Belous, Vitali Saladukha, and Siarhei Shvedau

Understanding GPS: Principles and Applications, Second Edition, Elliott D. Kaplan and Christopher J. Hegarty, editors

For further information on these and other Artech House titles, including out-of-print books available through our In-Print-Forever® (IPF®) program, contact:

Artech House
685 Canton Street
Norwood, MA 02062
Phone: 781-769-9750
Fax: 781-769-6334
e-mail: artech@artechhouse.com

Artech House
16 Sussex Street
London SW1V 4RW U.K.
Phone: +44 (0)171-973-8077
Fax: +44 (0)171-630-0166
e-mail: artech-uk@artechhouse.com

Find us on the World Wide Web at: www.artechhouse.com